T0100214

HOW THE VICTORIANS
TOOK US TO THE MOON

HOW THE VICTORIANS TOOK US TO THE MOON

The Story of the 19th-Century Innovators Who Forged our Future

Iwan Rhys Morus

PEGASUS BOOKS
NEW YORK LONDON

HOW THE VICTORIANS TOOK US TO THE MOON

Pegasus Books, Ltd.
148 West 37th Street, 13th Floor
New York, NY 10018

Copyright © 2022 by Iwan Rhys Morus

First Pegasus Books cloth edition December 2022

ISBN: 978-1-63936-260-8

10 9 8 7 6 5 4 3 2 1

Printed in the United States of America
Distributed by Simon & Schuster
www.pegasusbooks.com

ABOUT THE AUTHOR

Iwan Rhys Morus is professor of history at Aberystwyth University. He has published widely on the history of science, with titles including *Michael Faraday and the Electrical Century* (Icon, 2017), *Nikola Tesla and the Electrical Future* (Icon, 2019) and the *Oxford Illustrated History of Science*. He lives in Aberystwyth, Wales.

Acknowledgements

I have been thinking about the Victorians and their science for a very long time – for most of my career as a historian of science, in fact. I'm fascinated by Victorian culture because in so many ways it's still very close to us, but also often irreducibly alien. The origins of many of the ways we think and do things now lie in their century, even if we manifest them in some very different ways. This book is so long in the making, and I've had so many conversations along the way, that I cannot possibly thank everyone who has contributed to it in some way, but there are some I want to thank in particular. Conversations with Will Ashworth, Peter Bowler, Sarah Dry, Patricia Fara, Rob Iliffe, Bernie Lightman, Richard Noakes, Sam Robinson, Anne and Jim Secord, Charlotte Sleigh, Andrew Warwick and, of course, Simon Schaffer, at various times and in various places, have helped me enormously over the years as I tried to put my thoughts about the Victorian future into a more coherent framework. I'd like to thank Marina Benjamin for encouraging me to start the process of developing some of these ideas in aeon.co. Thanks also to my agent, Peter Tallack, for his enthusiasm and patience. I'm grateful to Andrew Furlow, Duncan Heath and James Lilford at Icon Books, for everything, and to Jo Walker for the fantastic cover design; and to

Claiborne Hancock and Nicole Maher at Pegasus, as well as Faceout Studio for the American cover.

Finally, my wife and best critic Amanda Rees has been, as always, wonderful. This is for you, Mandy.

Contents

Prologue
Inventing the Future

*Not that long ago, in a galaxy not
really all that far away ...*

It was 16 July 1909. There was a thunderous roar as His Majesty's spaceship *Victorious* rose imperiously into the blazing blue sky, a stately column of silver and gold balanced precariously on a tongue of fire. His Majesty himself, Edward VII, had travelled all the way to India's Deccan Plateau to see this latest triumph of scientific and technological ingenuity. Accompanying him were the prime minister, Herbert Asquith, the Royal Society's president, Sir Archibald Geikie, and the Royal Astronomical Society's newly elected president, David Gill, as well as the First Sea Lord, Admiral of the Fleet, Sir John Fisher. The prestigious gathering of notables only served to underline just how epoch-making the momentous occasion really was. It was the prelude to a pioneering journey of exploration unparalleled in history. Authors of scientific romances, such as the Frenchman Jules Verne or even H.G. Wells, had merely speculated about putting men on the Moon. Now, thanks to the combined expertise of the Empire's

1

engineers and men of science, it was really happening. This was no flight of fancy – it was taking place before their very eyes. In the tiny landing lighter *Deliverance*, perched on top of the huge rocket, three of His Majesty's most experienced naval officers were ready to take the Empire into space and claim the Moon for Britain.

The triumphant flight of HMS *Victorious* was the culmination of more than twenty years' determined effort by the leading men of science and engineering to conquer space and show to the world the superior reach and power of British technological ingenuity. The idea had first been mooted at the annual meeting of the British Association for the Advancement of Science in Bath during the summer of 1888. The society's president that year had been the eminent engineer Sir Frederick Bramwell, and during the public dinner that concluded the meeting, he had started speculating about just how far into space projectiles might be fired. He was interested in big guns, after all, and had delivered an address on the topic to the Birmingham and Midlands Institute just two years earlier.[1] While it soon became clear that no gun, however big, would be sufficient, someone suggested that something along the lines of a rocket of some kind might do the job. Gradually, the enterprise took shape. Lord Salisbury, the Tory prime minister at the time of the meeting in 1888, was a scientific man and was easily persuaded that sending men to the Moon would not only be a scientific triumph, but that it was absolutely essential for the good of the Empire that Britain should get there first. Imperial and industrial rivals might not yet have the resources to accomplish such a stupendous task, but they would one day. It was imperative that Britain should lay claim to the Moon and its resources before it fell into potentially hostile hands.

As the enterprise took shape, committees were formed to deliberate over the immense task ahead. Naval architects from the Royal Institution of Naval Architects, more used to designing dreadnoughts than rockets, debated competing plans for a space travelling vehicle. The British Association for the Advancement of Science (BAAS) and the Royal Society bickered over which institution should take the lead – although that issue was resolved with the establishment of the National Physical Laboratory in 1900. In 1879, the BAAS had deliberated whether it was economically feasible to construct the Analytical Engine that Charles Babbage had designed, but never built, in the 1840s. They had thought the cost prohibitive then, but now the machine was essential, and engineers struggled with the task of not only getting it made but made much smaller and able to work by electricity, not steam. Chemists experimented to find the most efficient fuel and electricians worked on the complex circuitry that would allow the crew to control the colossal space-flying machine. Resources from all over the Empire and beyond were poured into the attempt – the costs involved were astronomical. The three naval officers who would risk it all for the Empire were carefully selected and rigorously prepared – only the most self-disciplined men would be fit for the great adventure.

Four days after its successful launch into space on the tip of the *Victorious*, the *Deliverance* landed safely on the Moon's surface and for the first time in history, human feet stepped onto an alien world. The landing ground had been carefully selected – an apparently unobstructed area in the Sea of Tranquillity. When the three men stepped out of the *Deliverance* and stood on the Moon's surface, they were prepared for anything. They were armed, of course. There was a distinct possibility that this apparently lifeless surface might still contain life – the remnants, maybe, of some former civilisation that had degenerated and collapsed as the lunar atmosphere seeped away into space. If some degenerate life remained, then it might well be hostile. The selenauts came prepared to prospect for potential resources as well. Was there water, hidden in some crevices somewhere, for example? A supply of water would be essential if Britain were ever to establish a permanent station on the Moon to exploit what mineral resources might be there. But in many ways, the mission's main objective had already been achieved. The Union Flag now flew proudly over the Sea of Tranquillity, proclaiming to the world that the Moon belonged to Britain.

None of this really happened, of course, at least not in this universe. But there is still something compelling about this story. One reason for the contemporary popularity of steampunk, for example, is the sense that this fantasy of contemporary technology grafted onto the Victorian past is just teetering on the edge of reality.[2] We can believe in Victorians with steam-driven computers. And we can believe Victorians or Edwardians travelling to space in ways we can't really imagine of their predecessors. We can picture them belonging there, in ways that would be difficult to conceive a Puritan divine, or a Regency buck. One of the reasons it is easy to imagine Victorians on the Moon is that they imagined it themselves. The Moon seemed to be within the Victorians' grasp, teetering almost on the brink of reachability. Not only in the writings of those authors we still read today – Jules Verne or H.G. Wells, for example – but in the stories told by dozens of others, Victorian readers travelled to the Moon and beyond.[3] Writers that we have forgotten, such as George Griffith or Edwin Pallander, took their readers beyond the atmosphere, as well. There was a sense in which the Moon was almost familiar territory by the end of the nineteenth century, so often had the place been visited by scientific romancers.

Victorian writers were not the first to imagine going to the Moon, of course. The Bishop of Hereford, Francis Godwin, fantasised about travelling to the Moon in his *The Man in the Moone, or a Discourse of a Voyage Thither*, published posthumously in 1638. In it, he speculated that flying chariots might travel beyond the atmosphere and to the Moon, towed by a flock of geese. Inspired by the example, another English cleric, John Wilkins, later Warden of Wadham College, Oxford, and then Bishop of Chester, speculated in similar fashion in his *The Discovery of a World in the Moone*. Both clerics used their speculations about Moon travel as a way of popularising the latest astronomical ideas about the plurality of worlds – the view, based in theology, that not only must there be many worlds like ours out there, but that, like ours too, they must be inhabited. The key difference between stories like these and Victorian speculations is that Victorian writers really thought that travel to the Moon and beyond was within their grasp. Their science already possessed – or would soon possess – the means of getting there. It wasn't only scientific romancers that thought this. The year 1900 saw a flurry of popular speculation about what the world would be like at the end of the new century – and the end of the second millennium. Travel to the Moon was routinely cited as a technological feat that would have been accomplished by then.

A key reason for this confidence was that a new way of thinking about the future and its possibilities was emerging during this time – the way we think about the future now, in fact. New technologies, new ways of making knowledge and new visions about the future came together during the nineteenth century to create a new kind of world. Just 50 years earlier, most people assumed that the future would simply be an extension of the present. Nothing much would change. Another king might sit on the throne in a hundred years, but no one thought the world would turn into a completely different place. Forward-looking Victorians, on the other hand, were proud that they lived in an age of progress. It was what made them different. They congratulated themselves on the ways they were transforming the world around them, just as they prided themselves on having the self-discipline to turn dreams of the future into reality. They turned men of science and engineers into heroes. Samuel Smiles included many of their biographies in his 1859 book *Self-Help* – he even suggested that reading about the lives of such great men was as useful as reading the gospels (a truly shocking thing to say in mid-Victorian England).[4] The Victorian middle classes flocked to industrial and scientific exhibitions where they could see the future that science and technology would create taking shape before them. And if that were not enough, then they devoured scientific romances when they returned home. This book is about that transformation and the people who accomplished it, and how it produced our world today.

The very idea of progress was quite new and exciting at the beginning of the Victorian age.[5] A young John Stuart Mill, who would mature into liberal England's leading philosopher, wrote enthusiastically about the coming times. 'The conviction is already not far from being universal, that the times are pregnant with change; and that the nineteenth century will be known to posterity as the era of one of the greatest revolutions of which history has preserved the remembrance, in the human mind, and in the whole constitution of human society,' he said. 'It is felt that men are henceforth to be held together by new ties, and separated by new barriers; for the ancient bonds will no longer unite, nor the ancient boundaries confine.'[6] It was only in times of change, he thought, that people seriously considered the difference between the present and the past – and the future. Underpinning the idea of progress and change in society – that things can only get better – was a new understanding of change in nature. The world wasn't static any more. Unlike the old cosmos, for ever in equilibrium, the Victorian universe had a sense of direction.

Proponents of the nebular hypothesis – first suggested by Pierre-Simon Laplace – argued that the Solar System had not always been as it was since the creation of the world. It had begun as a cloud of dust and gas, floating in space, gradually coalescing into clumps of matter orbiting around a solid central mass. Over aeons of time, that central mass became the Sun, and the clumps of matter orbiting around it became the planets. The same process was still taking place elsewhere in the Universe, as new systems slowly coalesced out of the nebulae observed by William Herschel, or by Lord Rosse with his gigantic telescope, the Leviathan of Parsonstown, during the 1840s.[7] According to transformationists, those who believed in the idea of evolution, it was not just planets that had emerged from cosmic dust, but life as well, slowly working its way up the ladder of complexity towards humankind. Radical social thinkers clung to ideas like these as evidence that change needed to happen in society too – that progress was part of the proper order of things. After mid-century, Charles Darwin's ideas about evolution by means of natural selection demonstrated to Victorian minds that competition and the survival of the fittest were natural and entirely inevitable elements of progress too.[8]

There was a downside to progress, though. The new science of energy implied that the world could not last for ever. There had to come a time when progress stopped. It was a basic principle of the energy physics developed during the second half of the century that work could only be done when energy flowed from a hot body to a colder body. But that process made the hot body colder, and the cold body hotter, as well. Eventually, when everything in the Universe had arrived at the same temperature, no more work could be done, no more energy could be transformed. There could be no life and no progress. This was the heat death of the Universe.[9] At the same time, human progress carried the seeds of its own destruction.

More civilised societies coddled the unfit, so they bred and put natural selection into reverse. The very speed of modern life made people nervous and unbalanced. Society would degenerate. H.G. Wells played both with heat death and degeneration in *The Time Machine*, as his time traveller encountered the degenerate Eloi and Morlocks of the future as he travelled forwards towards the end of life itself.[10] Built into these scientific theories and romances was the recognition that the future would be different – that it was a strange new world that needed to be conquered and controlled.

In all sorts of ways, the Victorians were deeply invested in the future they were in the process of inventing, and in how it would come about. Theirs was going to be a technological future, produced by science and innovation. They could see the future being made in just this way all around them. New inventions, like the telegraph, the telephone and the radio, fed this vision of a future transformed by science. Right at the dawn of the Victorian age, satirists were already poking fun at the very notion of a future packed full of technological wonders. They pictured outlandish steam-driven chariots and baroque flying machines. Passengers were hurtled from one end of the Empire to the other through pneumatic tubes. But by mid-century, even as they lived in an increasingly steam-driven world, more people were dreaming of an electrical future. Increasingly, it almost seemed as if it were impossible to talk at all about electricity without invoking the role it would have in transforming the future. There would be electrical vehicles and electric power generated directly from the forces of nature. It would be electricity that powered the flying machines that the Victorians imagined filling the future's skies – and flight was a central feature of how the Victorian future was imagined.

So far, I have been using the term 'Victorian' in a fairly sloppy fashion. The Victorians who are the main protagonists of this book were of a very specific kind. They were overwhelmingly middle class, and even more overwhelmingly male. There is a reason why the little piece of fiction that I introduced at the beginning of this chapter sounds rather like something out of the *Boy's Own Paper*. It reads like that because the kinds of people who saw themselves as the makers of the Victorian future tended to see the world that way. They viewed the world around them – and the world that they were in the process of reinventing – through a very particular lens, and they saw it as belonging, and deservedly belonging, to people like themselves. There was something quite deliberate about the ways in which the protagonists in this book went about remodelling themselves and their institutions with an eye to the future. This was Victorian exceptionalism in action – they really thought they were different. Inventing what the future as a destination meant for people like themselves was part of the process of divorcing themselves from their own past. They prided themselves on possessing a self-discipline that their parents and grandparents had lacked – and it was that self-discipline that would make the future possible.

This book is about how and why the future was re-imagined by the Victorians and what went into that re-imagining. It sets out to try and understand what connects the Victorians' future to us – and to our own imagined futures. Our present may not look very much like the futures they imagined, but to a very large extent, the ways in which we extrapolate from our present to our future is very similar to the ways they did it. Like them, we make our futures out of bits and pieces of our present, and we see it as made by science and technology. Fictions matter as much as facts in the ways we imagine our future. Our future is made as much out of *Star Trek* as it is out of scientific actualities. That was how Victorian futures worked as well. They were made by Verne and Wells, and other authors we've forgotten, as much as by Michael Faraday and Isambard Kingdom

Brunel and countless forgotten experimenters and engineers. Just like ours, Victorian fantasies about the future offered ways of dealing with present-day dilemmas. But they described a destination, as well. They were where Victorians thought they were going, and they described the route for getting there.

Reforming science and its institutions was a vital element in inventing the future, and that is why this story starts with the battle for the soul of the Royal Society of London during the first half of the nineteenth century. The self-styled reformers who tried to take the society's reins from the old guard who had dominated it for much of the past half-century wanted to turn it into the embodiment of a new and more disciplined science – a science firmly directed at transforming the future. Led by Charles Babbage and John Herschel, they were convinced that only people like them had the right kinds of qualities to change science. They agreed with their enemies that science was best done by gentlemen, but they had very different views about just what gentlemen were meant to do to become men of science. As far as they were concerned, it was all about self-discipline. They thought their enemies were mere dilettantes – and corrupt dilettantes at that. Fellowship of the Royal Society should be granted on what they knew, not whom they knew. Out of the science wars and these bloody battles between ambitious men spanning three decades emerged a new understanding of science as a process that needed specialised knowledge and disciplined minds (men like the winners, in other words – in their view, at least).

The new generation of engineers were just as keen on discipline, and they thought that their practical know-how and entrepreneurial spirit was the key to the future. Men such as Marc Brunel and his son Isambard, or that other father–son duo George and Robert Stephenson, tunnelled under the Thames and criss-crossed the landscape with railways. They could do this – in their view – precisely because of their unparalleled practical experience of men and machines. Their ability to get things done was built into their own bodies through hard work and application. Newspapers turned them into heroes. Relationships between the self-made practical men and the gentlemen of science could be fraught, even though they were increasingly to be found in the same sorts of places, sitting in the same committee rooms. At the beginning of the Victorian age, they still embodied quite different ideas about discipline, for one thing, and were fighting for different visions of the future. During the course of the century, however, both groups came to share a common culture of accuracy and precision. Science and engineering became co-dependent. There were electrical engineers in the Royal Society and men of science in the Institution of Electrical Engineers. They shared a common cause in making their expertise count in the corridors of political power so that they could actively engage in future-making.

One of the places where science and technology looked more and more like the same thing was the exhibition. Men of science and men of engineering were equally keen on mounting spectacles of discovery and invention as a way of selling their visions of the future. They put their wares on show at places like London's Adelaide Gallery and the Royal Polytechnic Institution. The huge success of the Great Exhibition in 1851 inaugurated what many acknowledged was an Age of Exhibitions. People went to the Crystal Palace in Hyde Park, as they went to later exhibitions in Paris, Vienna, Philadelphia and Chicago, to see the future – and the raw materials out of which that future would be forged. It was there that the Victorian public learned how the future would be made and what it would look like. But progress was also the business of speculative writers who filled the pages of popular magazines like *Cassell's* and *Pearson's* with visions of the technological futures – dystopian and utopian – that these scientific and engineering men might produce. In the end, it was this confluence of expert and visionary that made the Victorian future – and still casts its shadow over how we imagine the future now. At international exhibitions and in the pages of popular magazines, the Victorians could see what the future would bring and who would make it.

There is no escaping the overwhelming masculinity of this world, or its thorough grounding in the business of empire. Even if men of science and engineering were agreed about nothing else, they were agreed that what they did was a man's affair. 'Men of science' was how the new breed of disciplined scientific gentlemen described themselves. The term 'scientist' had been invented to describe them in 1833, but it was rarely, if ever, used before the closing decades of the century. Even if it was hardly ever used, though, the fact that anyone thought it necessary suggests that some people, at least, understood that science had become something different now, and that there was a new breed of person doing it. The 'someone' in this case was William Whewell, polymath, mathematician and later Master of Trinity College, Cambridge. He derived the new word by analogy with 'artist': a 'scientist' would be someone who practised science in the same way that an 'artist' was someone who practised art – and it is worth remembering that, for Whewell, art meant work done by hand (so rather like engineering). In part, even though he was one of their generation, Whewell was trying to draw a line between himself (a philosopher) and the scientific reformers by coining a new term for them: it was a recognition that they were intent on doing something different with the world of science.[11]

Men of science and engineers were singular men, according to this new way of looking at things. Everything they did, everything they achieved, was down to them as individual movers of the world. When Samuel Smiles argued that the 'spirit of self-help is the root of all genuine growth in the individual', he might have been (and probably was) thinking about men like these.[12] Progress was the singular achievement of singular men: 'The worth and strength of a State depend far less upon the form of its institutions than upon the character of its men,' as far as Smiles was concerned, and 'national progress is the sum of individual industry, energy, and uprightness, as national decay is of individual idleness, selfishness, and vice.'[13] Even as men of science and engineers alike battled to transform their institutions and ways of collectively doing things, the focus of the stories they (and others) told about themselves were unremittingly about individual character. These were men who had made themselves. As far as they were concerned, it was Michael Faraday's 'conjunction of Poverty and Passion for Science', and his relentless drive for improvement that had made him the 'hero of chemistry', and not the careful training he had received at the hands of Humphry Davy, Europe's leading chemist.[14]

Anyone who scoured the pages of *Self-Help* looking for improving biographies of women who had successfully helped themselves would have been disappointed. There are no women in *Self-Help*. As far as Smiles, his readers and the men they were urged to emulate were concerned, individual self-improvement was an entirely masculine affair. The notion that a woman might be a role model, or even a potential reader, would have simply been inconceivable. Men of science were clear that their science was a man's domain. It required a cast of character that was distinctively masculine. Only men – and not even all men, but men of a very particular kind – had the right sort of mind for science. That the application of those minds actually depended on the hidden labour of armies of women was simply never acknowledged. If women had a place in science at all, it was as conveyors of knowledge made by men. So, Mary Somerville, for example, might write her *On the Connexion of the Physical Sciences* to huge acclaim from scientific gentlemen, or Arabella Buckley might charm children with her *Fairy-Land of Science*, but they weren't making anything new.[15] As far as the new breed of scientific and engineering men were concerned, not only was making the future men's work, it could not possibly be anything other than a masculine business.

The key, in that respect, was discipline. It was discipline, as far as many Victorians were concerned, that made the difference between women and men (or certain kinds of men). Men were, or were supposed to be, self-disciplined. They could keep themselves under control. Women, on the other hand, were too much at the mercy of their bodies to be capable of such discipline and therefore of making new knowledge. 'The instances of men in this country who, by dint of persevering application and energy, have raised themselves from the humblest ranks of industry to eminent positions of usefulness and influence in society, are indeed so numerous that they have long ceased to be regarded as exceptional,' said Smiles.[16] His biographer, John Tyndall, identified his iron discipline as the key to Faraday's pre-eminence in science. 'Underneath his sweetness and gentleness was the heat of a volcano,' Tyndall averred. Faraday 'was a man of excitable and fiery nature; but through high self-discipline he had converted the fire into a central glow and motive power of life, instead of permitting it to waste itself in useless passion.'[17] This was the kind of discipline that was needed to turn the world on its head and turn science into a tool for remaking the future in a Victorian image.

And that image was an imperial image. Smiles, for one, was perfectly clear about the relationship between the discipline of science and engineering and empire-building. It was the 'indomitable spirit of industry', of the kind that underpinned Victorian science and engineering, that had 'laid the foundations and built upon the industrial greatness of the empire, at home and in the colonies.'[18]

Victorian science and engineering were both made for empire and entirely its production. The future they were designed to generate was to be an imperial one. Throughout the century, it was the needs of empire that offered men of science some of their best arguments in favour of their disciplines. As the Welshman William Robert Grove put it: 'Why is England a great nation? Is it because her sons are brave? No, for so are the savage denizens of Polynesia: She is great because their bravery is fortified by discipline, and discipline is the offshoot of Science. Why is England a great nation? She is great because she excels in Agriculture, in Manufactures, in Commerce. What is Agriculture without Chemistry? What Manufactures without Mechanics? What Commerce without Navigation? What Navigation without Astronomy?'[19] The reformed science of the Victorians was honed to fulfil the requirements of relentless imperial expansion.

The discipline that was embodied in the science of accuracy and precision that flourished and grew during the nineteenth century was, as Grove observed, part of what made imperial governance possible. The physics of energy that reigned dominant throughout the second half of the century provided the scientific underpinnings of steam and telegraph – the technologies that governed the Empire. The institutions that were forged or reforged during the Victorian period – like the reformed Royal Society and new disciplinary bodies like the Astronomical Society or the Society of Telegraph Engineers – were unabashedly imperial institutions. The expertise and the expert knowledge made in those places circulated throughout the Empire and were essential to its maintenance – that's what they were for, after all. The Empire's resources provided the raw materials and the finance to build the Victorian age's great engineering and scientific achievements. Those resources and achievements alike were to be seen, admired and owned at the century's great exhibitions. And at those exhibitions the intimate link between dreams of empire and future dreams was forged and put on display. The kind of technological future that the Victorians imagined needed all the resources of empire to make it real.

The future that the Victorians invented was made up of the bits and pieces of their present. When electrical experimenters described their discoveries, for example, they described the future those discoveries would bring about, as well. Quite literally bringing the future into the Victorian middle-class home, Alfred Smee, electrical inventor and surgeon to the Bank of England, for example, told his readers how they would 'enter a room by a door having finger plates of the most costly device, made by the agency of the electric fluid'. The walls would be 'covered with engravings, printed from plates originally etched by galvanism', and at dinner 'the plates may have devices given by electrotype engravings, and his salt spoons gilt by the galvanic fluid'.[20] It seemed impossible to talk about electricity at all without talking about the future as well, and in this case, what was on offer was a future in which Victorian middle-class aspirations would be fully satisfied. Just as important in making this future were the scientific romancers who turned to the present's latest technologies to imagine new worlds in the future. They took men of science's promissory notes about the possibilities of their inventions and turned them into fictional reality. In popular magazines and scientific journals, the Victorian future slipped easily back and forth between fact and fantasy.

Victorians were particularly fascinated by what would fuel these new worlds. They knew what fuelled the present and knew that the supply of coal would not last for ever. Even as the Stephensons were putting their steam locomotive, the *Rocket*, through its paces at the Rainhill Trials in the autumn of 1829, satirists such as William Heath were drawing cartoons poking fun at the prospect of a steam-fuelled future. Within a decade of those trials, steam was already looking old-fashioned. From the late 1830s onwards, electricity was eclipsing steam in the public's imagination. Popular and technical journals were full of plans to build electric boats and electric locomotives. Experts argued over whether electricity would become more economic than steam power. The future pictured in scientific romances was overwhelmingly electrical, with electricity generated in limitless amounts from wind, sun or water. Nikola Tesla promised a future of wireless electricity everywhere (a future still being touted as late as the 1920s by Tesla promoters, such as the comics entrepreneur Hugo Gernsback). Meanwhile, committees of sober experts gathered at international exhibitions and set about establishing the standards and processes that really would make electricity the fuel of the future.

Invented by Charles Wheatstone and William Fothergill Cooke in 1837, the telegraph taught the Victorians just how transformative electricity could be. They were especially fascinated by the way it seemed to play with conventional notions of time and space. Information could travel at the speed of light and arrive at its destination quicker than the fastest means of human transport. When Alexander Graham Bell invented the telephone in 1876, the notion that a voice could be heard without the presence of the speaker seemed fantastic. Newspapers speculated that people would soon have telectroscopes – instruments for transmitting vision over long distances – in their parlours. George du Maurier had fun in *Punch* with a cartoon showing a comfortable Victorian pater and mater chatting to their children in the colonies. Telectroscopes were soon staples of scientific romance. Wireless telegraphy also spawned speculation of all kinds – about communication with other worlds, for example. Louis Pope Gratacap even imagined that the wireless might allow communication with the dead who had passed on to their next plane of existence on Mars. But these kinds of technologies were instruments of surveillance and oversight too.

The Victorians needed to find new ways of dealing with the avalanche of numbers they had created. As far back as the 1820s, Charles Babbage was already fantasising about calculating by steam. His Analytical Engine may never have been built, but the concept – as well as the mechanical calculators in common use by the end of the century – raised questions about the nature of intelligence and humanity. It was food for satire, too, though. As far as some critics were concerned, the notion that it would be possible to build an artificial human, indistinguishable from the real thing, was simply a joke. But what would it take to make a real automaton? Edward Page Mitchell speculated in the *New York Sun* about fitting Babbage's machine inside a human head to create a genius. Both E.E. Kellett and Alice Fuller imagined a perfect female automaton. Tesla fantasised about the art of telautomatics that would make war redundant, whereby self-acting weapons working by wireless would make conflict a matter of action at a distance and that deploying their destructive power would be unthinkable. Would machines in the future be able to process information as well as – or better than – mere humans?

Flying was the ultimate Victorian fantasy. Flying machines were frequent topics of discussion and speculation in engineering magazines. The engineer George Cayley, for example, was an assiduous promoter of powered flight. Models of his prototypes could be seen at the Adelaide Gallery and the Royal Polytechnic Institution, which he helped establish in the 1830s. Cayley wanted to claim for Britain 'the glory of being the first to establish the dry navigation of the universal ocean of the terrestrial atmosphere'. His ambition underlines the fact that from the beginning this was a fantasy about military and imperial power. Just as the British Navy ruled the waves and made the Empire possible, so inventors imagined how Britain might dominate the sky – or what the consequences might be of allowing some other power to do so. George Griffith's *The Outlaws of the Air* had anarchists terrorising Europe with flying machines. By the beginning of the Edwardian era, Britain's dreadnoughts were the last word in technological warfare. Those seemingly impregnable ironclads provided the model for what warships of the air might look like. What would happen to all that power if the fantasy of war in the air was ever realised? Tesla argued that automated flying machines powered by his wireless system would bring all war to an end. These were the technologies, real and fantastical, that Victorians imagined filling their futures, and their ambitious reach for the skies – and for the stars beyond – perfectly captures the hubris that powered their making of a science designed for world-changing.

In many ways, the future that the Victorians made is the future that we have inherited. By that I don't mean to suggest that they predicted our present (they didn't), but that we think about our future in the same way as the Victorians did – and that this way of imagining the future as a different country, to be arrived at through science and technology, was very much a Victorian invention. In this respect, we still imagine our current future according to a Victorian rule book. The Victorians' future, in fact and fiction, was generated to a large extent by extrapolating from the present. Telegraphs and telephones made it possible to imagine the telectroscope. Wireless communication across the sea made it possible to fantasise about communication with other worlds (or in Tesla's case, to speculate that he had already achieved it). Ironclad battleships sailing the seas made ironclads in the sky – or flying through space – more plausible. When Victorians read about these anticipated futures, they knew that they had seen and read about elements of them already. They had seen them at world fairs. They had read about them in magazines and newspapers. They knew a future made out of them was simply a matter of time.

So, as far as those Victorians who went to these exhibitions, or read about the latest scientific marvels in popular magazines, were concerned, their future was going to be scientifically generated. In some senses, that was what science was now supposed to be for. It was meant to be a tool for refashioning their world, and for refashioning their society, too. Creating futures was an essential aspect of the business of invention. Along with their understanding that the future would be one generated by science came the Victorians' view of who the makers of the future were. The future was going to be made by singular men, and it would be a future made by design for particular kinds of people – people like them, in other words. This is why understanding the Victorian invention of the future – of the way we now think about the future – is so important. As we've inherited the way we imagine and try to make our own future from the Victorians, we've also inherited some of the unthinking assumptions that came with that Victorian rule book. The best way of trying to think through those assumptions that still guide us is to understand where they came from in the first place, and how they were put to work.

Chapter 1

Science Wars

Sir Joseph Banks was dead. When he finally died at about 8am on 19 June 1820, the tyrannical old man had been president of the Royal Society of London for Improving Natural Knowledge for more than 40 years. Banks' lengthy reign at the pinnacle of England's premier scientific society had offered him a unique opportunity to shape science in his own image. The 'loss to Science by the demise of this excellent man and liberal patron will be long and severely felt', said the newspaper notices of his death.[1] The networks of influence and the power of patronage he had amassed during his office were certainly worth fighting over. Within days of his death, a fierce battle royal was already being waged over his legacy. The *Morning Post* reported a rumour that both Prince Augustus Frederick, Duke of Sussex (one of the many sons of the late George III, and brother of the soon-to-be-crowned George IV), and Prince Leopold of Saxe-Coburg (later king of Belgium) had been suggested as possible candidates for the presidency. It was, the paper thought, 'highly indecorous to the Royal Personages, disrespectful to the Royal Society, and disgraceful to the character which the Members of that

Learned Society ought to maintain', that anybody not possessing 'the scientific acquirements and general fitness of those whose character, talents, and acquaintance with the objects of the Royal Society, may designate as Candidates', should put their names forward like this.[2]

To understand why the vacant presidency of the Royal Society following Banks' death seemed such a ripe plum that even royal princes might think it was worth plucking, we must take a journey right to the heart of English science as it appeared during the early decades of the nineteenth century. Banks had worked hard during his presidency to make it into such a valuable prize. By 1820, the Royal Society's presidency was at the centre of a far-flung web of patronage and influence. As president, Banks had been assiduous in concentrating as much power as possible in his own hands. He and his cronies dominated not just the Royal Society, but the Board of Agriculture, Kew Gardens, the British Museum and the Board of Longitude, to name just a few of early nineteenth-century London's centres of cultural and intellectual power. Banks had close links to the court of George III too. He could use his ability to offer fellowships to the influential to bolster his own power. He could offer patronage, he could make introductions to the right quarters, he could offer funds. In short, Banks was the man who could make things happen in the world of English science. Anyone who wanted to get anything done in the scientific field had to make sure of Banks' support to achieve it.

The science war that engulfed the Royal Society for the next decade turned out to be a crucial one, with far more at stake than control over the society's purse strings or the power of patronage. This would become a battle for the soul of science. The men who had been waiting in the wings for an opportunity to stamp their own authority on the Royal Society had a very different vision from Banks and his cronies of what science was, how it should be ordered and what it could achieve. The campaign by these men to wrest control of the Royal Society from Banks' allies and fellow travellers and make it their own was the first of a protracted series of skirmishes spread out over the next two decades. And by the time the science wars were over, the Royal Society, and the vision of science it stood for, had changed irrevocably. To understand that transformation and what it meant for the future of science, we need to begin with Banks and to explore the uses he made of the Royal Society during his own long tenure at its head.

The story of Banks' success began in 1768 when he was appointed as the official naturalist to travel with Captain James Cook on his voyage to the south seas on HMS *Endeavour*.[3] He was the eldest son of a wealthy Lincolnshire landowning family and since his father died in 1761, while he was still a student at Christ Church, Oxford, Banks had control of the entire family fortune. His wealth and family connections made it easy for him to pursue his enthusiasm for botany. By this time, he had taken part in one voyage of exploration already. Two years earlier in 1866, he had sailed to Labrador on the HMS *Niger* captained by Sir Thomas Adams. It was his connection with John Montagu, Earl of Sandwich, First Lord of the Admiralty, that secured him his place on the *Niger*, accompanied by his Eton school friend Constantine John Phipps, who was already a navy lieutenant. Those connections also helped him to get his fellowship of the Royal Society in the same year while he was still in North America.

It was the success of this voyage that led to his appointment to the *Endeavour* two years later. That voyage would be a joint enterprise between the Admiralty and the Royal Society to observe the 1769 transit of Venus from Tahiti. It was a typical eighteenth-century collaboration, with both participants conscious that a successful observation would be both a national triumph and provide useful knowledge for navigation. The Royal Society's council asked in particular that 'Joseph Banks Esq., Fellow of the Society, a gentleman of large fortune, who is well versed in natural history', should embark on the voyage too.[4] The expedition made Banks. He came back with a shipload of unique specimens from the new continent that Cook discovered in the south seas. Their judicious distribution in the right places helped cement Banks' reputation and make him powerful friends in influential positions. In 1776, he bought 32 Soho Square and used his new London residence to host regular soirées and entertainments. By then he was already a member of the Royal Society's council. When Sir John Pringle, president of the Royal Society, resigned his position in 1778, Banks seemed the obvious successor.

When Banks took on the presidency, the Royal Society had been in existence for a little over a century. It was by design a society of gentlemen, for reasons both philosophical and pragmatic. It was important philosophically because the status as truth-tellers of the gentleman fellows was held to guarantee the knowledge they produced: it was all a matter of who could be trusted.[5] More pragmatically, the society's survival depended on its leaders being able to circulate smoothly through the upper echelons of English society. By the 1760s, the Royal Society was an integral part of a web of mutual patronage and influence that included departments of state such as the Admiralty, cultural institutions such as the British Museum and commercial enterprises such as the East India Company. This was the Royal Society that Banks inherited when he became president. During the following 42 years, Banks would work hard at maintaining and expanding that network of interests. Symbolically, in 1780 he physically moved the Royal Society from its old, cramped quarters in Crane Court to more luxurious apartments at Somerset House on the Strand, right at the heart of commercial, fashionable and political London.

Somerset House where the Royal Society moved in 1780.
Engraving from Thomas H. Shepherd, *London and its Environs
in the Nineteenth Century* (London: Jones & Co., 1828)

Under Banks' presidency, Somerset House was set to become
a vital centre for the making of useful and polite knowledge. Both
Banks himself and the Royal Society generally had extensive links of
patronage and mutual interest with the Admiralty. Under his direction, the Royal Society collaborated with the Admiralty in voyages
of collecting, exploring and exploiting natural resources in newly
discovered territories. Scientific men who wanted to make their
reputation through these sorts of activities needed to go through
Banks and the society to find opportunities for work. The case of
the botanist Robert Brown offers a good example. He had to lobby
Banks – and get others to lobby on his behalf – to get a place on
HMS *Investigator* as a naturalist.[6] In return, Banks expected to get
his share of the specimens Brown collected. William Bligh's ill-fated
voyage on HMS *Bounty* to collect breadfruit plants from Tahiti and
deliver them to the West Indies was another of the Banksian Royal

Society's collaborations with the Admiralty.[7] This was another example of how science could be made useful and turned to individual profit, as well as offering a way to make gentlemanly and professional reputations.

The East India Company was another important node in the web of influence surrounding the Royal Society. By the end of the eighteenth century, the company was already in many ways effectively an arm of the state and an important instrument of empire as well as a source of substantial commercial profit for its members.[8] Banks used his influence there to promote both his own and the Royal Society's concerns and offered advice as to how they might best exploit the natural resources of the territories they governed for economic gain. As the King's adviser on matters botanic, Banks could use the Royal Botanic Gardens at Kew as a repository for exotic flora brought back from voyages and expeditions of exploration. Kew was the centre for their redistribution as adornments for aristocratic gardens or as new crops for exploitation at home or in the colonies. It was Banks that suggested to the East India Company that the climate in some of the territories they controlled in north-east India was ideal for growing tea. In 1801, Humphry Davy carried out a series of experiments at the Royal Institution (another important node in Banks' web of influence) on the potential use of catechu (or *terra japonica*) in the tanning industry. He had acquired his supply of the plant from Banks, who in turn had acquired it through his connections with the East India Company. It should be quite clear from these connections whose interests Banksian knowledge was meant to serve.[9]

To his enemies – and Banks acquired many during the course of his long presidency – the system he had developed to control English science was corrupt, venal and self-serving. 'Those haters of light – the crafty, intriguing, corrupt, avaricious, cowardly, plundering, rapacious, soul-betraying, dirty-minded BATS' was how Thomas

Wakley, the firebrand founder of *The Lancet* described the leaders of the Royal Colleges of Physicians and of Surgeons in 1832, but the words could just as easily have been used by one of Banks' critics fifteen years earlier to describe him and his regime.[10] It is not surprising that the strident attacks on the Banksian model of polite and useful knowledge that increased in volume and intensity during the final years of his presidency coincided with an increase in equally strident attacks on establishment corruption more generally. They were all part and parcel of the same shameful state of affairs as far as hopeful reformers were concerned. Many of Banks' detractors were representatives of a new generation of men of science who had quite different views of how science should be done and what (and whom) it should be for. They detested the Banksian system while at the same time desperately wanting to get their hands on the power and patronage it represented. The war between them and the old guard became a war between competing visions of science, and between alternative futures.

The business of knowing

On Friday 13 November 1807, a group of gentlemen met at the Freemasons' Tavern on Great Queen Street near London's Covent Garden. It was not a particularly auspicious day for establishing a new scientific society, particularly since – as Humphry Davy pointed out – there were thirteen of them assembled there, to add to the bad luck.[11] Nevertheless, this turned out to be the first meeting of the Geological Society of London. The attendees may not have realised it, but it was a meeting that would set the scene for a different vision of science and its significance. Besides Davy, the flamboyant galvanic maestro from the Royal Institution, the group included the two Quaker brothers Richard and William Phillips, who were the sons of a London bookseller and printer and who both had interests in chemistry and mineralogy. Another of the founder members was their fellow Quaker, mentor and teacher William Allen, a practising chemist and pharmacist who, along with the Welshman Silvanus Bevan, ran an apothecary shop at Plough Court off Lombard Street. As a Quaker, Allen was a fervent abolitionist. The chemist William Haseldine Pepys was another attendee. The meeting was presided over by George Bellas Greenough, a lawyer and heir to a fortune made from the patent medicine trade, who would later be appointed as the society's first president.[12] With few exceptions – the most obvious being Davy himself – these were not the sorts of men who moved comfortably in the exalted circles occupied by Joseph Banks and his coterie at the Royal Society.

Whether or not the little gang of thirteen that gathered at the Freemasons' Tavern really intended their fledgling society to challenge Banks' iron grip on English science is unclear. It is perfectly

Freemasons' Tavern where both the Geological and
the Astronomical Societies held their first meetings.
Engraving, *European Magazine and London Review*, 1811, 59

clear, though, that that was how the Royal Society president soon came to understand it. Banks had already countenanced the establishment of some new scientific societies during his presidency. He had been happy to support the establishment of the Linnean Society in 1788, for example. After all, it catered for men who shared his interests in botany and collecting, and who showed no particular inclination to strike out independently in any case. He was rather more suspicious of the Geologicals, however, particularly as he came to understand that it was to be more than the mere dining club he had first thought it to be.[13] The problem was that the Geologicals did not seem willing to accept that their society would only operate as a satellite revolving around Banks' Royal Society. They also seemed to represent a different vision of what science was and how it should be organised. As far as Banks was concerned, the Geologicals were specialists (a term that acquired something like its modern meaning around this time) in a way that the Linneans, for example, were not. This meant that they were a lesser breed of philosopher. They were too focused on the details – too narrow-minded. They might have a fine grasp of those details, but they missed the grandeur of the big picture that was the real business of science.

That was not how the Geologicals saw it. From their perspective, the grandeur and the use of science lay precisely in those fine details. As Leonard Horner, one of the early joiners, described, they were concerned with the systematic collection of geological knowledge. 'Their object,' as Horner put it, required 'a minuteness of detail, and trouble, which they conceive would not be consistent with the present views of the Royal Society to undertake – yet they are aware that they are entering upon a branch of science, which is an important object of attention to the Royal Society and that very valuable communications on the subject frequently appear in the Transactions of the Society.'[14] This was not work for dilettantes or those with a superficial interest in the science. It could 'only be done by the united exertions of those who have attended to this subject, and who by their profession, their local advantages, or other circumstances, have been enabled to make important observations.' Horner was one of the foremost in his conviction that the Geologicals' future lay in making sure that only the properly proficient in their science should be permitted to join: 'We certainly ought to be more nice, and render the admission more difficult than perhaps it is at present,' he thought. The more Banks heard of this kind of view of how scientific societies should work the less he liked it. For him, it was a matter of who could be trusted with knowledge: people like him, or this new breed of specialist men.

Banks requested that he be made a member of the Geological Society, in particular, that he be made an honorary member, which was a category that was only meant for members who lived away from London and could not regularly attend meetings. The society voted to elect him to an ordinary membership, nevertheless. This certainly was an expression of independence. Things came to a head when it became clear that the Geological Society wanted to acquire their own premises to hold meetings and store collections of speci-mens. Davy – by now virtually acting as Banks' representative on the Geologicals' council – opposed this strongly. Banks himself, along with the wealthy collector Charles Greville, who had been appointed the society's patron, insisted that the 'President and several Fellows of the Royal Society consented to be enrolled in the Geological Society as an assistant Society without Funds, and its meetings lim-ited to a few dinners at a Tavern.'[15] In other words, their membership was on the understanding that the Geological Society was to be sub-servient to the Royal Society. Greville warned Greenough that if the Geologicals' plan went ahead, 'the Geological Society will become no longer an assistant, but a subverting Society, for Fellows of the

Royal Society bound to an ancient and respectable Incorporated Society would be bound to secede, and the great objects of National credit and of Science will risque being sacrificed to the Vanity or folly of individuals'.[16] There were indeed some resignations in the end – including Banks and Davy – but the Geological Society's leaders stood their ground in defence of their notion of specialist, disciplined knowledge.

When Banks was approached in 1809 by a group hoping to establish a Society for Animal Chemistry, the Geological experience led him to insist that his support would only be forthcoming if the society made its subservience to the Royal Society explicit. Specialist knowledge had to be reined in. The new society's council was to consist solely of fellows of the Royal Society, and it was agreed that 'all Discoveries and Improvements in Animal Chemistry, made by any of the Members, after they have received the approbation of this Society, be presented to the Royal Society'.[17] Banks' efforts to uphold the subservience of specialist scientists a decade later were rather less successful, however. This attempt at secession began at the Freemasons' Tavern too, when a group of fourteen gentlemen met on 12 January 1820 to establish the Astronomical Society. The list of those present included Charles Babbage, the son of a wealthy banker, and John Herschel, son of William Herschel, the discoverer of the planet Uranus. There were also Arthur and Francis Baily – two stockbrokers – as well as the actuary Benjamin Gompertz.[18] The group had met as they thought that the Royal Society was damaging the business of astronomy. If Banks was a representative of old money, these men were representatives of the new, and they wanted a new way of managing knowledge.

There can be no doubt that the Astronomicals were quite deliberately setting out to antagonise the *ancien régime* of English science. Establishing the Astronomical Society was the opening gambit in an attempt to force a change and promote an alternative vision of how knowledge should be produced and who should be trusted to produce it. Banks took no time in responding to the challenge. His first move was to persuade the mathematician Duke of Somerset, Edward Adolphus Seymour, to refuse the society's offer of its presidency. Francis Baily, newly appointed as the Astronomical Society's secretary, wrote to Babbage that if Banks 'casts the first stone, he must expect to be attacked in return ... Sir Joseph tells the Duke that our Society will be the ruin of the Royal Society: no mean compliment to us, but not very respectful to that learned body'.[19] Baily had indeed been the primary moving force behind the society's foundation. As John Herschel described in his memoir of his friend: 'It becomes impossible from this epoch to separate the Astronomical Society from astronomical science, in our estimate of his views and motives, or to avoid noticing the large and increasing devotion to its concerns of his time and thoughts'.[20] The society embodied what he and others like him thought science should be.

Baily had published what was in effect a manifesto for the new society a year before the meeting at the Freemasons' Tavern. It was 'much to be lamented', he argued, 'that in this country there is no association of scientific persons formed for the encouragement and improvement of Astronomy'.[21] It was a national tragedy that 'Astronomy, the most interesting and sublime of all the sciences (and, to our country, certainly the most useful) cannot claim the fostering aid of any society', he said. There were plenty of astronomers and plenty of observatories, but they were hamstrung by the lack of efficient coordination and opportunity for association. Individual observatories were all very well, but 'the utility of those establishments must be greatly circumscribed through the want of some mode of general communication among observers, by means of which their labours might be collected and registered; and thus rendered permanently useful. The formation of an ASTRONOMICAL SOCIETY would not only afford this advantage, but would in other respects be attended with the most beneficial consequences.' He dismissed the Royal Society itself in a footnote: 'The name of the Royal Society will naturally occur to the reader on this occasion: but that society was formed for the promotion and encouragement of science in general; and the subject of Astronomy appears to form but a small portion of its labours.'[22]

The Royal Society was not the only target for Baily. He also had the Board of Longitude – charged with overseeing efforts to improve ways of finding longitude at sea and packed with Banks' cronies since the 1818 Longitude Act – in his sights. The board, he strongly hinted, was not performing. As a result of its indolence, Britain was falling behind its continental competitors in the business of astronomy: 'The Board of Longitude in this country have now the power and the means of affording similar assistance by enlarging the original plan and design of the *Nautical Almanac,* and by assimilating it to those which are published at Paris, Berlin, Vienna, and other places: a measure which would tend to retrieve the character of the work, and redound to the honour of the country.'[23] The board already had the resources, they simply were not making proper use of what they had. Under Banks' regime, John Herschel later recalled that 'Mathematics were at the last gasp, and astronomy nearly so ... The chilling torpor of routine had begun to spread itself largely over all those branches of science which wanted the excitement of experimental research.'[24] Baily and his co-conspirators wanted to turn their science around by setting up an independent society and by wresting control of the Board of Longitude and its well-funded *Nautical Almanac* from the old guard. They wanted to instil some system and some discipline into their science.

Just like the founders of the Geological Society a decade or so earlier, the Astronomicals stood for a new vision of how science should be done and by whom. Discipline was to be the key to producing trustworthy knowledge – and the key to recognising a proper knowledge-maker. The Astronomicals had close links to the new commercial world of Regency London. The Baily brothers were stockbrokers. Gompertz and another founder, Henry Thomas Colebrooke, had worked as actuaries and had close links to the East India Company (Colebrooke's father was a former chairman of the company). The banker's son Charles Babbage was already dreaming about his Calculating Engine (he would announce his projected invention at a meeting of the Astronomical Society a few years later) and imagined it would be as useful for producing actuarial as astronomical tables.[25] All these men certainly saw similarities between the financial and scientific worlds and thought that science could profit from a dose of the discipline that made for prudent accounting. It was their possession of disciplined and specialist knowledge that was going to make them fit and trustworthy gentlemen of science. To achieve that end they needed not just to break free from the Royal Society, they needed to bring the Royal Society itself under control.

The banner of reform

Banks' death in the summer of 1820 gave the specialists exactly the opportunity they had been waiting for. With him out of the way, there was a real opportunity to do things differently at last. One of the first salvos in the war for the soul of science that would grind on for the next decade was fired in an article in the pages of the *Philosophical Magazine* just a few months after Banks passed away. Its author tore into Banks and his reputation with real venom. The late president was portrayed as a man who had been unfit by both attainment and character to hold the position he had occupied. He was pictured scurrying around the Royal Society's meeting room, bullying fellows into blackballing the nominations of candidates of whom he disapproved. 'To be sure Nicholson is a clever fellow,' they had him say of William Nicholson, late editor of the *Journal of Natural Philosophy* and one of the first to observe the decomposition of water by electricity, 'but you know he is only a sailor-boy turned schoolmaster; and we cannot, with any sort of propriety, admit such people among us'.[26] Banks' lavish entertainments came under scrutiny too. Those philosophers who were 'the glory of British science, would have had all their intrinsic excellency and all their distinguished celebrity, although Sir Joseph Banks had never gratuitously dispensed a single cup of tea'.[27]

This was a warning shot aimed at those who thought that Banks' successor as president should be a 'man of opulence' – someone like the royal princes whose names were already being bandied around as possible candidates, for example. The anonymous author, which was the mathematician and Astronomical Society founding member Olinthus Gregory, was clear that such a president was the last thing he and the new generation of specialist gentlemen wanted. The new president had to be one of them. While different factions were taking stock and choosing their candidates, the chemist and natural philosopher William Hyde Wollaston was quickly appointed as the Royal Society's temporary president. No one thought that he would be a suitable candidate in the long term – least of all Wollaston himself – but he was chosen to keep the seat warm while the warring camps decided who might best serve their interests as permanent president. The final choice was Humphry Davy. His backers hoped he would be recognised as a sufficiently eminent experimenter to please the reformers while still representing a continuation of the Banksian regime. In practice, he pleased no one. The reformers disliked him for his past associations and support for Banks, while many of the old guard simply regarded him as a jumped-up arriviste.[28]

Davy had first appeared on the metropolitan scene in 1801 when he was appointed lecturer in chemistry and assistant to the professor, Thomas Garnett, at the Royal Institution. Banks was one of Davy's patrons from the beginnings of his London career. Before his appointment at the Royal Institution, Davy had worked as laboratory assistant to the chemist Thomas Beddoes at his Pneumatic Institution in Bristol, where he helped investigate the therapeutic properties of different airs. Originally apprenticed to an apothecary in Penzance, he had come to Beddoes on the recommendation of fellow Cornishman Davies Giddy. In Bristol, Davy took laughing gas with Samuel Taylor Coleridge and Robert Southey as part of his experimental research on therapeutic airs. In London, he turned his attention to the newly invented voltaic pile (what would now be called an electric battery) and dazzled the fashionable Royal Institution lecture audience with spectacular electrical experiments. He hired Michael Faraday as his laboratory assistant, married a wealthy widow and received a baronetcy for his services. By 1820, he was lauded as England's (and, to the English, Europe's) greatest chemist. He was a key ally (and protégé) of Banks, but many of Banks' associates were also inclined to look down at his humble origins and ridicule the way he had risen through society by marrying money.[29] Sylvester Douglas, a prominent fellow and friend of Banks, called Davy 'a very little man'.[30]

As president of the Royal Society, Davy was in an impossible position, and his difficulties underscore the extent to which the situation was turning into a war between radically different cultures of knowledge. The men who saw themselves as the inheritors and defenders of Banks' intellectual legacy – men like Thomas Young, the secretary of the Board of Longitude responsible for publishing the *Nautical Almanac*, Davy's old patron Davies Gilbert (as Giddy now called himself) or even Davy himself to some degree – regarded knowledge primarily as an adornment for cultured gentlemen, albeit an adornment that could also offer them substantial economic benefit through schemes for agricultural improvement or colonial exploitation. That was not how the specialists saw it. As far as they were concerned, science was a vocation that demanded serious mental discipline. For them, science was certainly meant to be useful, but to be useful it needed system and it needed proper oversight. Men of science (and science was certainly regarded as something that required a masculine mind) needed to be self-disciplined if they wanted to discipline nature. In the wake of Banks' death and Davy's elevation to the presidency of the Royal Society, the specialists were quick to take advantage of the opportunities that a change of regime offered. They had the Board of Longitude and the superintendency of the *Nautical Almanac* in their sights, and they also wanted to reform the Royal Society itself to make it accountable to disciplined gentlemen like themselves.

It was the stockbroker Francis Baily, who saw himself very much as a business astronomer, that led the charge against the *Nautical Almanac* and the Board of Longitude. The board suffered from a lack of system, Baily challenged. Its members were dilettantes who lacked the firm and rigorous grounding in mathematics that was required to do their job properly. The *Almanac* had been established by the Astronomer Royal Nevil Maskelyne in 1767 to publish astronomical tables that could be used to help naval officers establish longitude at sea. Since 1818 it had been supervised by the Board of Longitude, which also controlled a budget of £4,000 a year to fund its publication. Under Thomas Young's dilatory supervision, Baily argued, money was being wasted and such funds could be used to forward the cause of astronomy if it was competently managed. Money was being thrown away that could pay for the production of socially useful specialist knowledge. It was 'a subject of much regret, among many scientific persons in this country, that the Nautical Almanac has not kept pace with those improvements in astronomy, and that spirit of inquiry, for which the present age is distinguished', according to Baily. England was falling behind: 'The honour of the country, and the interests of science demand that Great Britain should not be eclipsed by any of the minor states of Europe. And if the Nautical Almanac be designed also for an astronomical ephemeris – if it be intended for the observatory as well as for the quarter-deck – it ought to contain more than it now does.'[31]

The Board of Longitude's response to such criticism was revealing. Why should astronomy, asked Thomas Young, be singled out for largesse from the state? Paying for science was the prerogative of gentlemen who could afford it. Science was not something that gentlemen might do for pay. If other nations wanted to waste money on astronomical tables that was their business, and there was no reason why English astronomers should not help themselves to the work done by foreigners without needing to do the tedious calculation themselves either. This was a very different view from the specialists' one of what science was, why its practitioners should be valued and trusted and what knowledge was for. Baily was scathing about those who considered science 'as a mere article of commerce, which may be imported like hemp and tallow, and sold to the highest bidder – is there no merit in patronising the talents of our own country? – and are we for ever to be reproached (in this particular science) with using the manufactures of other nations, contented with supplying, in a few instances, the raw materials – the growth and produce of our own soil?'[32] It was outrageous that Young should dismiss the hard work of astronomical calculation as being simply a matter of curiosity. Knowledge and the labour that went into its making was being devalued.

One of the consequences of Humphry Davy's efforts as president of the Royal Society to please both hostile camps was that representatives of the specialist societies found themselves sitting on the society's council. John Herschel was one of the first Astronomicals to find himself on Davy's council. He was followed by Charles Babbage, James South – a former surgeon who had married enough money to allow him to devote all his time to astronomy – and Francis Baily himself. Davy did his best to support them in their efforts to push for reform of the *Nautical Almanac* too, though that was an uphill struggle in the face of Thomas Young's unwillingness to either compromise or relinquish his lucrative position as its superintendent. Crucially, though, the Astronomicals and their Geological allies on Davy's council were in a position to mount a coup. Things came to a head in 1827 when the council set up a committee 'to consider the best means of limiting the number of members admitted into the Society, and to make such suggestions on that subject as may seem to them conducive to the welfare of the Society'.[33] This sounded innocuous enough. In fact, it was dynamite. With the exception of two remnants from the old Banksian order – Davies Gilbert and John George Children – the committee was packed with new men. They wanted to change the rules so that in future only men who shared their vision of science could become Royal Society fellows.

By reducing the number of fellows and making sure that 'every vacancy would become an object of competition among persons of acknowledged merit', the reformers were proposing to stack the deck in favour of people like them.[34] No longer would presidents be able to curry favour among the powerful and build up networks of influence by offering fellowships to their patrons. Fellowships would be competitive, and the competition would favour the new breed of specialist committed to disciplined knowledge for the public good. Fellowship would be the hallmark of trust. That was what was meant to happen, and the report the committee presented to the council a few weeks later made concrete proposals for achieving the reformers' radical aims. The fellowship would be capped at 400 members and only four new fellows a year would be appointed until their numbers had been reduced accordingly from the 650 or so at that time. Instead, however, before the end of the year, the increasingly ill and disillusioned Davy resigned the presidency and after furious infighting his former patron Davies Gilbert was elevated in his place. Gilbert promptly kicked the reform proposals into the long grass, where they remained. It was a powerful reminder of the power still wielded by Banks' allies even several years after his death.

The setback left the reformers in disarray, and it was a few years until they were able to rally again. In November 1830, when Gilbert announced that he would be relinquishing the presidency at the annual meeting on St Andrew's Day, and the Duke of Sussex was put forward as his replacement, the specialists responded with fury. They promptly put up John Herschel as an alternative candidate, setting the stage for a showdown between the specialists' vision of science and the Banksian tradition. A notice in *The Times* set the tone. 'We know nothing of the disputes which agitate that learned body,' the paper said loftily (and mendaciously), 'but if knowledge and not rank should carry the day, Mr Herschel is a man of great science, in the highest department of science – Astronomy. It has always, and truly been said, that there is no royal road to learning: why should there be any royal road to the highest honours which can be conferred on learning?'[35] More than 60 fellows signed a notice asserting that 'Mr Herschel, by his varied and profound knowledge and high personal character, is eminently qualified to fill the office of President, and that his appointment to the chair of the Society would be peculiarly acceptable to men of science in this and foreign countries.'[36] Francis Baily's name was the first on the list.

It did not take long for the gloves to come off. A letter in *The Times* laid out the battleground. It was obvious, said the anonymous writer, that 'none but a person of some rank or political importance can hope to direct the scientific patronage of Government always into the proper channels, or be able to transact business with the public offices in a certain and expeditious manner'. That was what Joseph Banks had excelled at doing and that was what the Royal Society needed again, not the services of a mere specialist, according to the letter. It was obvious that such a man should have 'an ample fortune and an establishment of proportionate magnitude' as well as an 'elevated station to command respect, courteous manners to soothe and persuade, and an entire freedom from all suspicion of a predilection for any particular science, as a pledge that no partialities will be felt, and no preferences be shown'. It was a shame that the son of a man who owed his fortune to the King (Herschel, in other words) should be standing in the way of one of the late monarch's sons, the writer suggested.[37] This was a radically different view from the specialists' one of what the presidency of the Royal Society required of its incumbent. Another correspondent was clear on the same grounds that a 'scientific' president was the last thing the Royal Society needed. On election day the room was packed. Herschel lost by eight votes and the Duke of Sussex was named president of the Royal Society.[38]

Working knowledge

Astonishingly, in the immediate aftermath of his victory, some of the duke's supporters tried to pretend that there had really been no contest and that Sussex had gained the presidency of the Royal Society by acclamation – the vote had been for the election of the new council and since Herschel was not on that list, he could not be a candidate for the presidency, they said. They were fooling no one, of course. It was 'an absurd explanation' said *The Age* newspaper.[39] *The Times* was blunt in their condemnation: 'The first scientific establishment in the empire has obtained a Prince, and missed a Philosopher, for its President.'[40] The prince himself tried to smooth things over. He had 'heard, with the greatest regret, of the existence of very serious differences in the Society', he said, and would 'use his utmost efforts to put an end to those unhappy differences, and to restore harmony and the most cordial co-operation among all the Fellows of the Society'.[41] It was a forlorn hope. Babbage, for one, had already announced his own manifesto for change in his vitriol-laden *Reflections on the Decline of Science in England*, published just a few months previously, which effectively threatened to replace the Royal Society with a new state-sponsored academy for specialists.

In between broadsides condemning the Royal Society for its corruption and the incompetence of its officers, Babbage offered a vision of science supported by the state along the same lines as the French Academy of Sciences. By making itself open to any man with the means and influence to get himself elected as a fellow, the Royal Society had lost the trust of government and public alike. The fellowship was no longer an indicator of trustworthy knowledge. The 'pursuit of science does not, in England, constitute a distinct profession, as it does in many other countries', he complained.[42] This meant both that practitioners lacked status and that the public lacked a good way of knowing whom to trust when it came to knowledge. The specialist sciences required 'such unremitting devotion of time, that few who have not spent years in their study can judge of the relative knowledge of those that pursue them'. As a result, 'the public, and even ... men of sound sense and discernment, can scarcely find means to distinguish between the possessors of knowledge, in the present day, merely elementary, and those whose acquirements are of the highest order'.[43] Few of his readers would have shared Babbage's admiration of the French way of doing things – many of them thought the Academy of Sciences was just as corrupt as the

Royal Society – but his diagnosis of the problem was shared by many. They agreed that the Royal Society needed to be pummelled into a new shape that would make it fit for the future.

In an appendix to his outburst, Babbage reprinted an account he had published of a meeting he had attended in Berlin in 1828. He thought that the 'great congress of philosophers' open to all who professed science, offered a model for an organisation that could stand up to the Royal Society and offer a space for specialists to take their organised and disciplined knowledge to the public.[44] When a group of gentlemen met at York in 1831 to establish the British Association for the Advancement of Science, Babbage, for one, was quick to see the possibilities the new organisation offered. Since the late eighteenth century, local scientific societies (literary and philosophical institutions) had been proliferating in the provinces and particularly in the new industrial towns and cities. The idea behind

The British Association for the Advancement of
Science at its Swansea meeting in 1848.
Sketch by John Weir Padley

the BAAS was to find a way of bringing them all together on a yearly basis. Its annual meetings – always in a different location but never in the metropolis – offered an opportunity for local scientific talent to meet the metropolitan gentlemen of science. Meetings were organised into sections, each one devoted to a different specialism (chemistry, geology, physical science and so on). Huge crowds gathered for these annual scientific jamborees and their proceedings were widely reported in the local and national press. They were a shop window for the new and disciplined way of doing things.[45]

The BAAS and its sections provides just one example of the extent to which something more like the specialists' view of how scientific work should be organised was starting to gain some traction by the 1830s. Ironically, perhaps, another example is the Royal Society's joint project with the Board of Longitude to improve the quality of optical glass.[46] Instigated by Davy while president, on the grounds that 'the present state of the glass manufactured for optical purposes was extremely imperfect, and required some public intervention', a committee was established that ultimately set up a sub-committee to carry out some practical experiments. That committee consisted of the optical instrument maker George Dollond, Davy's laboratory assistant Michael Faraday and the leading light in the Astronomical Society John Herschel. Faraday would be responsible for making samples of different kinds of glass, Dollond would grind the glass into lenses and Herschel would test the glass's optical properties. This was to be careful, disciplined and systematic work for a very specific purpose and a specific public benefit. The experimenters were chosen because they were trained practitioners who had the necessary skills and knowledge for the task. It was to be an exercise in producing useful knowledge the specialist way – though it was also part of an attempt by Davy to broaden the scope of the Board of Longitude's activities.

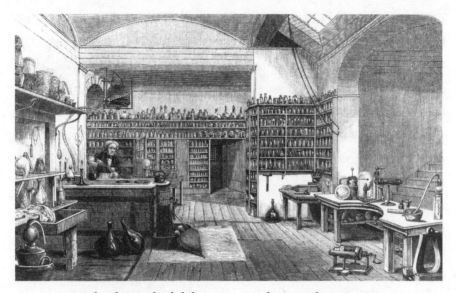

Michael Faraday's laboratory at the Royal Institution.
Engraving from Henry Bence Jones, *The Life and Letters of Faraday*
(London: Longman, Green & Co., 1970)

The glass project did not long survive Davy's death in 1829 – and neither did the Board of Longitude. It was abolished in 1828 although Young continued to supervise the coveted *Nautical Almanac* until his death in 1829. The concern was then taken over by the Royal Observatory until, in 1832, William Stratford, secretary of the Astronomical Society, was appointed as superintendent of the Nautical Almanac Office. By then it was the Astronomicals, rather than the Royal Society, to whom the Admiralty turned for advice on such matters. Just a few years previously, the Admiralty had commissioned a report from the Astronomical Society on the *Nautical Almanac's* future, and they had responded with an extensive list of suggestions for reform. Stratford's task was to implement them.[47] Wholesale reform continued a few years later when George Biddell Airy replaced John Pond – another former protégé of Banks who had occupied the role since 1811 – as Astronomer Royal. Airy had already made a considerable name for himself as director of the Cambridge

University Observatory where he had already introduced some of the measures to systematise the process of observation and calculation that he would adopt at the Royal Observatory in Greenwich. As Astronomer Royal, Airy proceeded to reform the Royal Observatory root and branch.

Under Airy's direction, the Royal Observatory was transformed into an astronomical factory.[48] A few years before Airy's appointment, Charles Babbage had written *On the Economy of Machinery and Manufactures*, extolling the virtues of the division of labour in factory work. The great advantage of the division of labour, he argued, was that you only needed to pay for the amount of skill needed for a particular task, and no more. This applied to mental labour as well – and the labour of calculating tables, such as those produced in observatories like Greenwich.[49] This was the system that Airy introduced. The tedious business of calculation was left to computers (the term then referred to humans rather than machines, usually young men with limited arithmetical skills who therefore worked for relatively little pay) and the checking and application was left to the highly skilled specialists like Airy and his assistants. This was the sort of system for generating reliable, trustworthy knowledge that the Astronomicals had dreamed about. It was with the aim of improving this system even further, as we shall see later, that Babbage would develop his Calculating Engine.

In the same year as Airy's appointment to the Royal Observatory, Henry De la Beche was appointed head of the newly established Geological Survey. While Airy's job was to systematically survey the heavens, De la Beche's was to inaugurate a systematic survey of subterranean Britain. The Geological Survey began as a division of the Ordnance Survey, which had its origins in the need for systematic mapping of the British coastline during the war with revolutionary France in the 1790s. What De la Beche wanted was a systematic map of Britain's geology. It would be funded by the state and would be an exemplar of what systematic and specialist scientific work could deliver for the national economic interest and the advance of trustworthy knowledge. Proper geological maps generated by trained and trustworthy specialists would lay bare the country's subterranean resources. De la Beche's friend George Bellas Greenough, president of the Geological Society, called on the society's fellows to 'rejoice in the complete success which has attended that first attempt of that honourable Board [of Ordnance] to exalt the character of English topography by rendering it at once more scientific and very much more useful to the country at large.'[50] Increasingly, it was this new breed of specialist and vocationally committed men of science who were becoming the people who could be trusted to make their knowledge useful. They sat on royal commissions or were called in as authoritative witnesses. To give just one example, both Michael Faraday and Charles Lyell were called to give evidence at the inquest following the Haswell Colliery disaster in 1844 – something that would have been unlikely just a decade earlier.[51]

So, by the 1830s, despite the fact that the Royal Society remained dominated by the old guard who had inherited Banks' mantle, and his view of science as polite knowledge for the leisured classes, the new breed of specialists were becoming increasingly powerful. Their claim to be the sole legitimate producers and disseminators of trustworthy knowledge was gaining traction in the right places. Their calls for the reform of the institutions of knowledge during the 1820s resonated with wider political calls for the reform of ancient and corrupt political institutions that led eventually to the 1832 Reform Act. In the aftermath of political reform, the specialists (who tended to be liberal in politics too) were the ones who found friends most easily in the corridors of power. Their view of knowledge as something useful in itself, but also useful for its capacity to generate social and economic progress, was increasingly appealing. It found its most powerful expression in John Herschel's *Preliminary Discourse on the Study of Natural Philosophy*, published in 1831 as the introductory volume of the Cabinet Cyclopaedia, which aimed to bring the fruits of specialist knowledge to the middle-class reading public.[52]

Herschel's tract laid out what knowledge should look like from the specialists' point of view and what it was for. Proper knowledge was useful in 'showing us how to avoid attempting impossibilities'; it was useful in 'securing us from important mistakes in attempting what is, in itself, possible, by means either inadequate, or actually opposed, to the end in view'; in 'enabling us to accomplish our ends in the easiest, shortest, most economical, and most effectual manner'; and finally, in 'inducing us to attempt, and enabling us to accomplish, objects which, but for such knowledge, we should never have thought of undertaking'.[53] But it could only achieve this if it was done in the right way by the right kind of people. Knowledge was meant to be transparent. It 'should be divested, as far as possible, of artificial difficulties, and stripped of all such technicalities as tend to place it in the light of a craft and a mystery, inaccessible without a kind of apprenticeship'. Its producers needed to avoid the temptation to take 'pride in particular short cuts and mysteries known only to adepts; to surprise and astonish by results, but conceal processes'. The mental discipline and the processes that produced reliable knowledge had to be visible so that people could see that the result could be trusted.[54]

The Royal Society remained unfinished business, though, and the reformers' defeat in 1830 still rankled. The Duke of Sussex continued as president until 1838, when he was replaced by another (if slightly less exalted) noble lord, the second Marquess of Northampton, Spencer Compton. He had a little more claim than his royal predecessor to scientific credentials – he had even been president of the Geological Society between 1820 and 1822, but the main reason he was given the Royal Society presidency was still his peerage. The real power behind the throne, though, was Peter Mark Roget (of later Thesaurus fame). Roget had been one of the society's secretaries since 1827 and was a skilled and ruthless back room politician. Discontent rumbled on – even at the meeting that elected Northampton as president, there was an attempt (quashed by Roget) to curtail presidential patronage by limiting the term of office. In 1846, a gang of reformers rallied for another attempt to take control of the society, led by the geologist Leonard Horner, a veteran of the failed campaign to make Herschel president, who persuaded the society's council to appoint a committee to examine the Royal Society's charter, with a mandate to examine ways in which it might be amended. 'Its present state is not wholesome,' Michael Faraday complained, adding wistfully: 'I do wish for better times.'[55]

Roget duly tried to hijack the charter committee, persuading it to recommend a raft of minor and anodyne changes to the society's process for electing fellows. The reformers retaliated by getting William Robert Grove – the recipient of that wistful letter from Faraday, as it happens – appointed to the committee. Grove was a rising star on the metropolitan scientific scene. Born to an affluent Swansea family, he had first made a name for himself in scientific circles with some brilliant electrochemical research, inventing a powerful new kind of battery that would soon prove its value in the growing telegraph industry, as well as a philosophical curiosity: the gas battery – 'a beautiful instance of the correlation of natural forces', as he described it.[56] Thanks to his research, he was made a fellow of the Royal Society in 1840, and a few months later was appointed professor of experimental philosophy at the London Institution, putting him on a par with Faraday in the panoply of London science. Crucially – and this is what made him such an important asset to the reformers – he was a lawyer. He had entered Lincoln's Inn following the completion of his studies at Brasenose College, Oxford, and had been called to the Bar in 1835.[57]

Grove was also thoroughly disenchanted with the state of science. In 1843 he had published a blistering attack on the state of 'physical science in England' in the pages of *Blackwood's Magazine*. The Royal Society's corruption and inertia had eaten away at the fabric of science, and the new specialist societies were just as bad, he argued. These ill-disciplined institutions 'do harm by the cliquery they generate, collecting little knots of little men, no individual of whom can stand his own ground, but a group of whom, by leaning hard together, can, and do, exercise a most pernicious influence'. By 'seeking petty gain and class celebrity, they exert their joint stock brains to convert science into pounds, shillings and pence; and when they have managed to poke one foot upon the ladder of notoriety, use the other to kick furiously at the poor aspirants who attempted to follow them'. The solution to this undisciplined state was to remodel the Royal Society as an institution above them all, 'accessible only to men of high distinction who would be thus constituted the oligarchs of science'.[58] That was what Grove, at least, wanted to achieve by way of reform when he joined the charter committee.

With Grove onside, the committee promptly threw out the list of lukewarm reforms that Roget had persuaded them to endorse. Instead, they opted for something far more ambitious – imposing a strict limit on the number of fellows who could be elected each year and making sure that it was the council, not the president, who would choose who was on that roll of successful candidates. Yet again, Roget tried to suppress the proposals put forward by the committee, proposing to the society's council that discussion of the proposals should be indefinitely adjourned. The gambit failed, and the council duly noted its approval of 'the two recommendations in the report of the Charter Committee; viz. that the number of Fellows to be elected in any one year do not exceed fifteen, and that the Council do recommend to the Fellows the most eligible Candidates'.[59] Northampton and Roget were incandescent – and both promptly announced their resignations. Northampton, in his final speech to the fellows as president, publicly disassociated himself from the society in its new form. 'As I was one of those who entertained considerable doubts of its prudence and expediency,' he huffed about the new state of things, 'I cannot claim any praise should it prove advantageous to the Society, nor must I be considered responsible in case of failure.'[60]

Grove and the reformers were cock-a-hoop. They prudently made sure that Northampton's successor was yet another noble lord, though, to placate the conservatives – but Lord Rosse was a noble lord who shared their views about the new way of doing things. They also helped establish the Parliamentary Committee of the British Association for the Advancement of Science in 1849 to lobby the government on scientific matters. That committee was chaired by another scientific peer, Lord Wrottesley, who would go on in due course to succeed Rosse as president of the Royal Society. To make sure that things continued to go their way, the reformers established a club – the Philosophical Club – charged with the task of nudging the Royal Society in the right direction. Ostensibly, at least, the new club's remit was 'to promote as much as possible the scientific objects of the Royal Society'. In reality, the aim was to have a forum for getting things done behind the scenes, and to make sure their vision of disciplined science prevailed. As Edward Forbes put it to Grove, they wanted to teach their fellow men of science 'to look at the Philosophical as a sort of higher council or guardian angel of them all'.[61]

The triumphant reformers viewed their processes for making knowledge as indispensable in every field. Herschel's manifesto for the production of reliable and trustworthy knowledge about nature was meant to be a model for all knowledge and a blueprint for how science could transform the future. The mental and moral discipline required to do science was essential in all areas of enquiry. The man of science was the exemplar of the impartial but committed searcher for truth. The knowledge produced by these gentlemen of science was meant to be public and useful. But the public had to trust the gentlemen who produced it too, if that knowledge was to do any good. The fact that science was meant to be transparent did not mean that just anyone could do it. It might no longer be the preserve of aristocratic dilettantes with their private collections of botanical specimens from far-flung places, but the new gentlemen of science were clear that knowledge-making was their monopoly. That was how the future would be secured. As we shall see in the next chapter, though, there were real tensions between this view of transparent and useful knowledge and the claims of practical men that they were the ones who really understood the world and could reinvent the future.

Chapter 2

Practical Men

On 10 November 1827, a group of gentlemen sat down to enjoy their dinner, deep beneath the muddy waters of the Thames near Rotherhithe in London. They were assembled in the Thames Tunnel that Marc Brunel and his son Isambard were in the process of boring between Rotherhithe, on the south bank of the river, and Wapping, on the north. A party of 40 or so distinguished guests assembled for the occasion, seated in one of the tunnel's two parallel arcades. In the other arcade were gathered about 120 of the tunnel's skilled workforce – the experienced miners and bricklayers, the foremen and overseers – who had laboured for months and years in the stinking mud under the river to carve out the Brunels' ambitious project. As they were ushered through the tunnel and sat down to eat, the assembled company was serenaded by the Coldstream Guards Band. The band was there as a special favour to Richard Beamish, the younger Brunel's assistant, and himself a former officer of the guards. The tunnel was illuminated by gaslight for the occasion, the floor was carpeted and the walls festooned with swathes of crimson velvet. It was an extraordinary occasion, even by the extravagant standards of the 1820s.

As the guests descended into the depths, the 'band struck up a favourite air from *Der Freischütz*'. As they entered the tunnel itself, the '*coup d'oeil* of the festive board was more like enchantment than reality; upon a nearer approach, however, the company found it no unsubstantial pageant, but an excellent and well-appointed dinner'. As the King was toasted, along with the Duke of York, the Duke of Clarence and the navy and the Duke of Wellington and the army, the toaster observed that 'the health of our gracious Sovereign had been drunk in every clime; but hitherto never under that river which carried over their heads the wealth of nations into the wealthiest metropolis in the world'. While the toasts were going on, the workmen had been conducting their own revels. At the end of the evening, 'a proposal was made by them, through their foreman, which will appear novel to such of our readers as are unacquainted with mining operations'. A 'pick-axe and spade were produced, and held up as the symbols of their craft, over which they requested three times three, in honour of the same; a request that was complied with in shouts, which for some minutes resounded with surprising effect throughout the whole length of the subterraneous, or rather sub-aqueous edifice'.[1]

Everyone sitting around those banqueting tables under the Thames understood what a powerful symbol the Thames Tunnel was. It was a typically bravura piece of early nineteenth-century engineering – ambitious and gargantuan in scope. The possibility of connecting the two banks of the Thames via a tunnel had been mooted since the final decades of the previous century. The canal builder Ralph Dodd had proposed such a tunnel in 1798, running from Gravesend to Tilbury with a view to speed up the deployment of troops to the coast if the French invaded. Dodd, who had made a name for himself as a mining engineer in the north of England, assured potential backers that 'however novel the idea may appear to some, it is both practicable, and pregnant with utility'.[2] He even attracted enough financial backing to start digging before repeated floodings brought things to a halt. A few years later, the Cornish mining engineer Richard Trevithick attempted the feat, moving the proposed tunnel to a shorter route between Rotherhithe and Limehouse. He brought in miners from his native Cornwall and started digging too, but flooding again prevented much progress. It was not until the French émigré engineer Marc Brunel came along with his proposal for a tunnel between Rotherhithe and Wapping in 1823 that things really got properly under way.[3]

Growing up in Hacqueville in France, Brunel's ability as a draughtsman led him first towards a career in the French Navy. By the early 1790s, though, his royalist sympathies were making life in France increasingly dangerous and he fled to America, armed with an American passport, begged from a sympathetic consul. There he started a new career as a surveyor and engineer, building fortifications in New York, among other things, and a factory for mass-producing cannonballs, as the city's chief engineer. In 1799, he crossed the Atlantic again, to Britain, with a letter of introduction from Alexander Hamilton, the commanding general of the US Army, to the First Lord of the Admiralty. Brunel thought he had come up with a solution to a problem that had been causing the Admiralty increasing difficulties in their war with revolutionary France: how to quickly and efficiently produce the wooden pulley blocks that were vital components of fighting sailing ships. The blocks were made by hand and quickly wore out, and the navy needed them by the thousand. Brunel eventually succeeded in persuading the First Lord that he really did know what he was talking about and was ensconced at the naval yards in Portsmouth, working with Samuel Bentham, brother of Jeremy Bentham, founder of utilitarianism and creator of the panopticon.

Brunel vastly improved the efficiency of block production by mechanising the process.[4] Designed by Brunel and manufactured by Henry Maudslay, who was celebrated as a maker of precision machinery, the block-making machines were marvels of engineering. It took some time, but by 1806 there were 43 of the devices operating at the Portsmouth dockyards, producing blocks in the tens of thousands. Brunel claimed that thanks to his machinery, six men could now do the work of 60. As Brunel's tunnel beneath the Thames would be in due course, the Portsmouth block-making machines were a major attraction for sightseers. Brunel's invention was hailed as an example of the wonders of automated manufacturing, and visitors flocked to the dockyards to see it in action. Brunel's machinery did not make him rich, but it did make him famous. His name had become one to conjure with in the world of engineering speculation. When he announced plans for a new and potentially lucrative project, he was a man who would be listened to. And by the beginning of the 1820s, Brunel had a new system in mind for tunnelling under the Thames and finishing the job that Dodd and Trevithick had started.

Grand projects like the Thames Tunnel were deliberately ambitious in their scope because they were quite explicitly intended to inaugurate a new way of doing things. They were meant to show how the moral courage, superior skill and sheer vision of practical men could bully nature into submission. These were the men who could, quite literally, reshape the landscape and the nation's future. Brunel – and his soon-to-be far more famous son, Isambard – had a secret weapon at his disposal that he believed would allow him to succeed where others had fallen foul of the oozing, stinking layers of mud beneath the river. This was the 'Shield', as they called the crucial piece of equipment that they hoped would keep the tunnel safe from becoming inundated. It was a simple enough idea – Brunel claimed it first came to him when watching shipworms burrow into the

wooden planks of ships at the Portsmouth naval yards while designing his new block-making equipment. The Shield was a huge frame
divided into a number of cells in which individual miners would
crouch, digging away, pushing the frame slowly forward as they
cleared the muck in front of them. If one cell started to leak water,
it could be easily blocked off before the entire tunnel was flooded.
Like Brunel's block making machinery, the invention was built
for him by Henry Maudslay.

The Shield was just the kind of simple idea that captured the
public imagination. It was this talent for devising clever innovations
that was meant to make the new breed of ambitious engineer stand
out. Newspapers made much of this new example of 'the romantic
projects of an enterprising people', and the key role of the Shield in
carrying it out.[5] They offered their readers detailed descriptions of
how the contraption would allow the miners to dig in safety. Each

Men working on the Shield digging the Thames Tunnel.
Engraving from the *Illustrated London News*

digger in his cell removed one of the boards that separated him from the riverbed, dug a cavity a few inches deep, replaced the board and repeated the process with the next one. Once each digger had excavated to the same distance, the whole Shield was moved forward and the process repeated. Behind the Shield waited squads of bricklayers, ready to brick up the tunnel's earth walls as they were revealed by the Shield's slow progress. It was the Shield that made the Thames Tunnel project into an enterprise that investors were willing to back. As Richard Beamish, Brunel junior's assistant, later himself the Tunnel's resident engineer, and ultimately Marc Brunel's biographer, put it, 40 years later: 'As the mode of effecting this gigantic excavation began to be understood, and the unrivalled mechanical capability of the projector was remembered, a feeling of confidence spread through the scientific and practical mind of the country, which resulted in the formation of a company to carry out the design.'[6]

This was the projector as hero – and heroic even when things went wrong. On 18 May 1827, after a few troubled weeks, the Shield broke and Thames water poured into the tunnel. Within seconds the works were flooded and the workers rushed towards the ladders that led to the surface and safety. 'The workmen fled towards the shaft in the greatest terror,' reported *The Times*. 'Mr Brunel, jun.' on seeing that one of them had fallen behind, 'immediately descended to his aid, and succeeded in rescuing him from his perilous situation, at the moment that his strength was almost exhausted.'[7] Other metropolitan newspapers carried much the same story.[8] A few days later they reported how 'Messrs. Brunel, sen. and jun., descended in the diving-bell' beneath the Thames to assess the damage, and how 'Mr Brunel, jun., tied a rope around his waist, and was lowered from the bell on the shield of the Tunnel.'[9] Brunel senior, the papers reported, had already assured the board of directors that he had 'adopted means to remedy the evil, and remove the water', and was 'confident that the work will, in a short time, be resumed, and proceed as usual'.[10] These were not men who recognised failure.

In due course, Brunel junior would be celebrated as one of his generation's greatest and most ambitious engineers. When he went back to save that drowning miner in the rapidly flooding tunnel, he was only 21 years old but was already taking on much of the day-to-day responsibility of overseeing the work. It was not going to be an easy task. In fact, it took four months – and the death of one of the workmen clearing the tunnel – before digging recommenced. It was to mark the resumption of work that the banquet beneath the river was organised. It was an emphatic declaration of defiance and determination. Determination was certainly needed. The tunnel flooded again in January 1828, this time injuring Isambard Brunel and killing several more workers. *The Times* reported that it was 'a source of satisfaction to all concerned, that Mr Brunel was on the spot, and that all that could be done for the safety of the poor sufferers was done, at great risk to himself.'[11] Work on the tunnel dragged on throughout the 1830s, and there were frequent inundations. The Brunels had to use all their considerable charisma to persuade shareholders to keep up the flow of money. There were appeals to government for assistance for such a groundbreaking and ambitious undertaking – and the projectors did receive £270,000 from the public purse.

The Thames Tunnel was in every way an engineering spectacle. Very soon after work on the tunnel started, paying visitors were welcomed to the works. This was not unusual, as the example of Marc Brunel's block-making machinery at Portsmouth showed. Factories and engineering works often attracted visitors during the first half of the century – they were tangible symbols of the coming future. The curious flocked to see a new world being made. But the Thames Tunnel took spectacle to a new level. On the practical side, the shillings paid by the curious to descend under the river and see the Shield in action were the only source of cash to keep the investors happy. There had even been a party of spectators in the tunnel only a few hours before it flooded in 1827, despite Marc Brunel's presentiments – 'May it not be when the arch is full of visitors!' he worried about the prospect of flooding just a few days earlier. 'I attended Lady Raffles and party to the frames,' (meaning the Shield) he recorded on 18 May 1827, 'most uneasy all the while, as if I had a presentiment, not so much of the approaching catastrophe to the extent it has occurred, but of what might result from the misbehaviour of some of the men, as was the case when the Irish labourers ran away from the pumps and the stage.'[12]

Once it was fully open to pedestrian traffic at last, the tunnel quickly gained its place as a London landmark. The opening ceremony that took place on Saturday 25 March 1843 was an astonishing spectacle. 'Another wonder has been added to the many of which London can boast,' crowed the *Morning Chronicle*, 'another triumph has been achieved by British enterprize, genius, and perseverance.' Charles Babbage was there, and Michael Faraday, in a parade of dignitaries that filed down into the tunnel, serenaded by the band of the Scots Fusilier Guards and accompanied by cheers 'which rung and reverberated through the subterraneous passage, and was echoed by the multitude without.'[13] Medals were struck to commemorate the great occasion. By March the following year, more than 2 million people had paid a penny to stroll beneath the river. Fairs and soirées were held there. It was a place for spectacles as well as being a spectacle in its own right. At the Annual Fancy Fair, people could marvel at moving panoramas and electrical displays as well as a model of the Shield itself. Visitors could buy souvenirs of their descent. As a commercial enterprise, it soon became clear that investors would never get their money back, but the tunnel's place as a spectacle – a vision of an engineering future – was assured.

The Thames Tunnel was a monument to the ambitions of practical men. But it was also a site for conflict over the question of who the practical men really were. Whose labour – and whose talents – underpinned it? For the Brunels, the answer was obvious. Marc Brunel designed the tunnel and his son oversaw the fulfilment of his father's ambition. The tunnel was testament to the genius of engineers like them who possessed the vision and the determination to make it real. Their workers may have felt differently, nonetheless. Work beneath the Thames was punctuated by strikes as the men who toiled beneath the river fought for a wage that recognised both the risks that they ran and the skills that they carried with them. The few weeks before the first flooding had seen strikes, and others followed. These workers were highly skilled miners and tunnellers – and they thought very highly of themselves. They had been hand-picked precisely because of their consummate knowledge of their craft. They knew that without their experience of the underground world – the idiosyncrasies of earth and mud and rock – there would be no tunnel. These were men who – like the Brunels – thought of themselves as being practical. Their hard-earned skills were an essential part of who they were. Highly skilled workers like these saw themselves as the aristocracy of the labour force, in the vanguard of the march into the future, but they stood for a very different vision of where credit for great engineering works should belong.

Mechanics' culture

On 30 August 1823, just as plans for the Thames Tunnel were gaining traction, a new magazine was published. The *Mechanics' Magazine* was created to offer a voice to the voiceless. There was 'no periodical publication', they said, 'of which that numerous and important

Frontispiece of the first issue
of the *Mechanics' Magazine*, 1823.

portion of the community, the Mechanics or Artisans, including all who are operatively employed in our Arts and Manufactures, can say, "This is ours, and for us."' They were sure that any of them who had 'a just sense of his own importance in the scale of society, must wish to see this want supplied'.[14] Coming as it did, less than a decade after the end of the war with revolutionary France, this was radical stuff. In the face of a rapidly changing world of work, as well as a repressive political climate, the magazine was part of an overt attempt to provide a forum where men who laboured with their hands could stand up for their rights and work to improve themselves. It was no accident that the magazine's first article was a biography of that quintessential working man made good, James Watt, of steam engine fame. It was chosen precisely because Watt's 'good fortune may encourage, and his perseverance instruct the present and all future generations of mechanics'.[15]

At the end of their first year, the editors patted themselves on the back, that 'so rapid was the success which attended our endeavours, that we can only venture to ascribe it to the extreme desire which prevailed for something of the kind, however imperfect or faulty might be the manner of its execution'.[16] They were aiming these words directly at the kind of skilled worker who would soon be labouring beneath the river on the Thames Tunnel (and they duly published an account of Brunel's ambitious plans). The radicalism might be well disguised, but this really was radical stuff, and the editors, Thomas Hodgskin and Joseph Clinton Robertson, knew exactly what they were doing. Both men were deeply involved in radical politics. Hodgskin was a former midshipman who had taught himself political economy at sea, and by the beginning of the 1820s he was scraping a living as a journalist (he wrote for the *Morning Chronicle* among other newspapers) and lecturer. He was a friend of the tailor turned radical Francis Place. When he took on the editorship of the

Mechanics' Magazine, he was already developing the alternative vision of political economy that would make him notorious.[17] He drew on traditional mechanics' views about the rights they had over the tools they used in their craft to assert their moral ownership of the machines that would make the future.[18]

Robertson was the son of a disgraced Edinburgh minister sent packing from Scotland for conducting illegal marriages. Moving in radical circles, like Hodgskin, whom he had met while still living in Edinburgh, Robertson was a patent agent and campaigner to reform patent laws. These laws were widely believed to discriminate against working mechanics by making it prohibitively expensive to patent their own innovations and allowing unscrupulous masters to pilfer their ideas. This was exactly the kind of issue that would keep the magazine in business – defending the defenceless craftsman against those who would deny them their proper dues. Under the pseudonym Sholto Percy he was one of the authors of *The Percy Anecdotes* – a kind of bluffers' guide to literary culture. Both men were clear that the *Mechanics' Magazine* was going to be a vehicle not just for promoting the interests of mechanics and artisans, but for redefining who they were. The new artisan would be self-made, and their carefully nurtured skills, and the instinctive knowledge they possessed about the machines they worked with, were what would make them the real makers of tomorrow.

Artisans – men (and they were almost invariably men) who had been trained in a particular craft – regarded themselves as being the vanguard of labour. Typically, they had served apprenticeships of seven years, during which they learned the specific skills needed to carry out their trade. At the end of their apprenticeships, they were expected to produce a masterpiece – a piece of work that demonstrated that they were able to carry out the requirements of their craft. So, a blacksmith might be expected to produce a piece of worked metal, a cabinet maker an item of furniture or a book-binder a bound book. If the masterpiece was deemed acceptable, the apprentice would become a journeyman and would be able to practise their trade independently. Some would go on to become masters themselves, running their own workshops and taking on apprentices of their own. All these men understood that they possessed a property of skill. They owned their skills just as they owned the tools of their trade. It was their skill that made them independent. It was also what gave them a sense of themselves and their own value. Being practical men was deeply ingrained in their selfhood.

It was not just artisans who understood their own value. From the 1780s until well into the nineteenth century, skilled workers were forbidden from emigrating from British territories.[19] Ports were policed to make sure that they did not leave. Government agents across Europe and the United States reported on British artisans working overseas. These men's bodies were valuable to the state because of the practical skills they carried with them. It was their value that could make them dangerous though. Because they knew their worth – and because of their defining sense of manly independence – they were difficult to control. That was one of the virtues of self-acting machinery according to its promoters. Andrew Ure (the Pindar of the factory system according to Karl Marx) even argued that 'wherever a process requires peculiar dexterity and steadiness of hand, it is withdrawn completely from the cunning workman, who is prone to irregularities of many kinds, and it is placed in charge of a peculiar mechanism, so self-regulating, that even a child may superintend it'.[20] It was in the face of this kind of challenge, not just to their economic power, but their very sense of who they were, that many artisans and their political spokesmen – people like Hodgskin and Robertson – wanted to assert that they were the real practical men, and that, morally, they were the proper owners of the machines that depended on their skills.

It was in the face of these kinds of challenges to the manly independence of practical men that the *Mechanics' Magazine* was established. They were also at the heart of the magazine's campaign to establish a London Mechanics' Institute. Institutes like this – aimed at providing a scientific education to working men – were already being created across the country. There was one established in Liverpool earlier in 1823, for example. But as far as Hodgskin and Robertson were concerned, in particular, manly independence was going to be at the very heart of the London institution. It would be established by mechanics and for mechanics: 'The education of a free people, like their property, will always be directed most beneficially for them when it is in their own hands.' Only by helping themselves could mechanics hope to retain their dignity and their autonomy: 'Men had better be without education – properly so called, for nature herself teaches us many valuable truths – than be educated by their rulers; for then education is but the mere breaking in of the steer to the yoke; the mere discipline of a hunting dog, which, by dint of severity, is made to forego the strongest impulse of his nature, and instead of devouring his prey, to hasten with it to the feet of his master.'[21] This was radical self-help, and it was meant to ensure that mechanics were properly recognised – by themselves as well as others – as the real practical men. As Hodgskin would later argue, artisans and mechanics 'were no longer mere workers with edged tools but had pressed the great powers of nature into their service. The mechanic in the gas works was a chemist of considerable skill.'[22]

A meeting was held in the Crown & Anchor tavern to further the campaign – ironically, since one of the motives for establishing such institutes was to move radical politics away from rowdy pubs and into more respectable spaces. Speaking there, the veteran radical William Cobbett, editor of the *Political Register* and author of *Rural Rides*, was emphatic about the need for manly independence:

'If they allowed other management to interfere, men would soon be found who would put the mechanics to one side, and make use of them only as tools.'[23] The London Mechanics' Institute was duly founded on this wave of enthusiasm – but the bitter quarrel that soon erupted between Robertson and the man chosen to be the institute's president, George Birkbeck, cast a revealing light on the tensions simmering beneath the surface. Robertson was already warning his readers that if 'the working classes, the mechanics of this metropolis, should at any time suffer this Institution to be perverted from its original purpose, their instruction in the principles of the mechanics' arts, and in other branches of useful knowledge, the fault will be entirely their own.'[24] As a professor of natural philosophy in Glasgow, Birkbeck had made a name for himself giving free lectures to the city's mechanics. As such, he was widely regarded as being a leading figure in the drive towards self-improvement. Moving to London, Birkbeck oscillated between scientific, financial and radical circles. He was a powerful and influential patron for the new institution – but patronage was the last thing that Robertson wanted.

Robertson used the pages of the *Mechanics' Magazine* to lambast Birkbeck for turning the institution into his own personal fiefdom, subverting the 'manly independence' he wanted to promote. The principle of independence was being entirely undermined by Birkbeck's behaviour: 'The Managers have proceeded on a system of begging for gratuitous instruction, wholly inconsistent with that spirit of independence which breathed in the first resolutions of the Society, and, in our humble opinion, most injurious to the real interests of the working classes.'[25] What Robertson and his allies had wanted was an institution that embodied their view of who the practical men really were, and how they should be organised for their own improvement. What they got was an institution governed by liberally minded philanthropists who wanted to impose a rather different sense of who they were and where they belonged. It was a battle for ownership of the future. Theirs was a vision of the social place that a new breed of practical men should occupy that drew heavily on customary artisanal notions of independence, mutual responsibility and collective ownership of skills and status. It was a notion of expertise as something that belonged to the group rather than to a single gifted individual.

This clash of cultures could have real consequences. As we shall see later, for example, Charles Babbage's ambitious project to build a Calculating Engine came to grief precisely because of disputes between him and Joseph Clement, the highly skilled craftsman he had hired to make its components. As far as Babbage was concerned, the Calculating Engine and the tools needed to build it belonged to him because they were products of his invention. Clement, in the tradition of his trade, thought they belonged to him. In much the same sort of way, Brunel saw the Thames Tunnel as uniquely his because it was his inventive genius that had conceived the Shield that was essential to build it. Like Clement, the miners who did the actual toiling in the Shield beneath the Thames had a different understanding of things. As far as they were concerned, what mattered was the skill and labour that went into making things, and those things came from being part of a collective. What Hodgskin and Robertson were trying to do in the *Mechanics' Magazine* was turn that sense of collectivity in a new direction. Their readers were going to be the practical men of the future. This is what they meant when they celebrated the 'classlessness' of the *Mechanics' Magazine*'s readership. It was all about the collective good. As Robertson put it, it 'may be safely questioned, whether there is any other instance of so many persons of so many different classes of society being brought thus to co-operate together, for their mutual improvement; the learned instructing the unlearned; the theorist assisting the practical man, and the practical man the theorist; and most of them preferring, manifestly, to every consideration of self-interest or ambition, the pure satisfaction which arises from the consciousness of doing good for their fellow-men.'[26]

The engineers

Hodgskin and Robertson might have imagined a future in which practical expertise was common property — if only the common property of a tightly restricted group of skilful men who thought of themselves as the elite of workers – but that certainly was not the future that engineers like Marc and Isambard Brunel saw coming. As far as they were concerned, invention belonged to them uniquely: it was their property. As Isambard Kingdom Brunel's son remarked of his father's relationship with his grandfather: 'He was indebted to him, not only for the inheritance of many natural gifts, and for a professional education such as few have been able to procure, but also for a bright example of the cultivation of those habits of forethought and perseverance, which alone can ensure the successful accomplishment of great designs.'[27] That kind of story of self-help and self-belief was fundamental to the ways in which these men saw themselves. It was also fundamental to how others saw them. Engineers were the heroes of the Victorian age, responsible for the bridges, canals and railways that transformed the landscape, and the steamships that linked up the Empire. The Victorians rarely had anything good to say about the men who laboured to build these things. The miners and the navvies who laid rails, dug tunnels or hacked through rock to make cuttings were objects of deep suspicion for the Victorian middle classes. They adored the engineers who designed and oversaw them though.

When Samuel Smiles published *Self-Help* in 1859 (the same year that Charles Darwin published *On the Origin of Species*), inventors were prominent in his pantheon of exemplary self-made men.

Smiles had made a name for himself telling earnest young men in mutual improvement societies about the best ways of bettering themselves. The book was a best-seller and turned Smiles into a celebrity. He drew on evangelical religion as well as the tradition of manly independence to promote the idea that self-discipline and perseverance were the keys to success – and engineers were the best examples. Shockingly, he told his readers that the biographies of great engineers were 'almost equivalent to Gospels – teaching high living, high thinking, and energetic action for their own and the world's good'.[28] The distinctive characteristic of men like this was that they had made themselves – and made themselves through hard work and self-discipline. James Watt was a great engineer, said Smiles, not just because he was a great thinker, but because he had kept his nose to the grindstone. It might well have been true that many 'men of his time knew more than Watt, but none laboured so assiduously as he did to turn all that he did know to useful practical purposes'.[29] Being an engineer meant being a particular kind of character.

When Smiles wrote *Self-Help*, he had already written a biography of the engineer George Stephenson. Born to a dissenting Edinburgh family, trained and apprenticed as a surgeon, by the 1840s, Smiles was re-inventing himself as a writer and lecturer. In 1838, he was appointed editor of the radical *Leeds Times*. As editor he campaigned for radical causes and dabbled in Chartism. It was this peculiar blend of evangelical dissent and political radicalism that bred self-help. Evangelicalism – very much on the rise during the first half of the nineteenth century – placed a particular emphasis on hard work and self-discipline as signs of grace.[30] Political radicalism valued the manly independence of the artisan. As Smiles himself noted, *Self-Help* had its origins in a lecture he delivered in 1845 to a Mutual Improvement Society (a kind of informal mechanics' institution).

Smiles was a friend and colleague of Stephenson senior – he worked as his assistant secretary while he oversaw the building of the Leeds and Thirsk Railway. He thought Stephenson the epitome of the self-made engineer – bluff, no-nonsense and straightforward. After Stephenson's death, he persuaded his son to let him write his father's biography. Following *Self-Help*'s success, Smiles made a career from engineering biographies. The five-volume *Lives of the Engineers* published a few years later further cemented the engineer's image as a self-disciplined, self-made and unambiguously masculine figure. Engineering flair was bred in the bone. It needed character, grit and determination.

The wonders these men wrought on the landscape – and on the Victorian imagination – seemed arrogant and glorious in their scope. Thomas Telford's Pontcysyllte Aqueduct, for example, glowered over the Dee Valley near Llangollen, carrying the Ellesmere Canal across the river. Apprenticed as a stonemason, Telford had built a reputation for himself as a builder of roads and bridges in his native Scotland during the closing decades of the eighteenth century. Moving south to further his engineering ambitions, Telford worked as a mason on Somerset House, home of the Royal Society. In Shrewsbury, he was hired by William Pulteney, one of the wealthiest men in Britain thanks to the profits of slavery, to work on repairs to Shrewsbury Castle. To Smiles' approval, he was obsessed with improving himself, as well. 'Now you know,' he wrote to a friend, 'that I am chemistry mad; and if I were near you, I would make you promise to communicate any information on the subject that you thought would be of service to your friend, especially about calcareous matters and the mode of forming the best composition for building with, as well above as below water.'[31] He carried on building roads and bridges too, and it was his success as a bridge-maker that led to his work on the Ellesmere Canal and Pontcysyllte.

The Pontcysyllte Aqueduct.
Engraving from Samuel Smiles, *Lives of the Engineers*
(London: John Murray, 1861)

Built of brick and cast iron, the Pontcysyllte Aqueduct was an entirely deliberate evocation of imperial Rome on the part of its builders. It was meant to underline contemporary imperial ambitions as well. As Telford put it, Pontcysyllte 'added a striking feature to the beautiful vale of Llangollen, where formerly was the fastness of Owen Glendower, but which, now cleared of its entangled woods contains a useful line of intercourse between England and Ireland; and the water drawn from the once sacred Devon furnishes the means of distributing prosperity over the adjacent land of the Saxons'.[32] He could have said much the same of his other masterpiece, the Menai Suspension

Bridge, the largest in the world, with its sixteen huge chains bearing the road across the strait a hundred feet above the water. It was built to hasten the passage of mail between London and the port of Holyhead on the western tip of Anglesey – and on to Ireland and Dublin: a vital artery of imperial power. Telford meant constructions like these to be deliberately hyperbolic. They were far more than a means of getting from place to place; they were statements of engineering might. They were promissory notes of mortar, bricks and iron for the future that men like Telford would build for Britain.

The railways that sliced through the landscape courtesy of the Stephensons carried a similar message. George Stephenson had worked his way up from employment as an engineman and then a brakesman in a colliery – he only learned to read at eighteen. In his late twenties he was made engine-wright at Killingworth Colliery and taught himself the secrets of steam engines. There in 1814 he built his first steam locomotive – the *Blücher*. His subsequent rise was spectacular. He built his first railway in 1820. In 1821, he was hired to survey and construct the Stockton and Darlington Railway. By then he had been joined by his son Robert. Stephenson junior had been apprenticed to the mining engineer Nicholas Wood before breaking his apprenticeship to work with his father. He spent a few months studying at Edinburgh University before heading for South America to work as a mining engineer in Colombia. He returned to build the *Rocket* and work with his father on the Liverpool and Manchester Railway. George Stephenson was certainly proud of himself as an entirely self-made man. He embodied everything that an engineer should be, and all his success was the product of his own labour and his own dogged force of character. Robert, clearly, could not be quite so self-made.

The 1830s and 1840s were the decades of heroic railway building – and men like the Stephensons were the powers that generated that frenetic expansion. That, at least, is how they saw themselves. Investors flocked to make their fortunes, betting their futures on the back of engineering men, and railway mania followed railway mania in rapid succession. With the Liverpool and Manchester Railway completed, the Stephensons moved on to build the London and Birmingham. This was going to be 'the most extensive and magnificent Railroad in Great Britain'. 'The mind is lost in amazement as it contemplates so vast and splendid a structure in all its bearings, in its yet undeveloped powers, and in its future results and influences upon the resources of a great country', according to one commentator.[33] Tellingly, its building was compared to the building of the Great Wall of China and the Egyptian Pyramids. In terms of materials used and resources needed, it put those ancient monuments in the shade. In fact, it seemed clear that 'such a work as this could only have been undertaken in a country abounding with capital and possessing engineering talent of the highest order'.[34] It was the capacity of men like the Stephensons to think big, to forge ahead and to oversee resources on an unprecedented scale that seemed to their contemporaries to make them appear almost superhuman. That was why Smiles saw them as such paragons of self-help.

The Brunels – and Brunel junior in particular – inspired similar adulation. The Thames Tunnel was the last of Marc Brunel's engineering labours, but it was the first in a long series of labours for his son. Only a few years after that banquet under the Thames, Brunel junior's design won the competition for a bridge to span the gorge at Clifton on the outskirts of Bristol. It took more than 30 years for it to be built. He was soon appointed the engineer for the Great Western Railway Company and started work westwards from London to the sea. As with the Stephensons' drive from London to Birmingham, the Great Western's engineer had to re-form the landscape, building bridges and carving out cuttings and tunnels.

Challenged by geologists over the safety of his diggings, he could retort 'that I ought now to possess a more thorough and practical knowledge of this particular rock and its defects, and the best mode of remedying them, than even you yourself, with your immeasurably greater scientific knowledge of rocks generally'.[35] Brunel's ground-breaking railway heroics offered a way of showing off the practical man's hard-won knowledge of the world, gained through intimate interaction with its idiosyncrasies. It was another way of showing the extent to which his superior engineering know-how really was Brunel's own personal property.

Brunel's dream of the Great Western was about more than just a railway, though. The way west did not have to stop at Bristol and the sea. Even as the railway was just a plan on paper, Brunel asked: 'Why not make it longer, and have a steamboat go from Bristol to New York, and call it the "Great Western".'[36] This was Victorian engineering at its brashest and most ambitious. What started (maybe) as a joke, turned into reality. The SS *Great Western*, too, was suitably gargantuan – though not on the scale of the steamships Brunel produced later in life. It was (for a while) the largest steamship built – Brunel was convinced that scale was the solution to steamship efficiency. His next steamship, the SS *Great Britain*, was even bigger and at 322 feet in length she was more than 100 feet longer than any of her rivals. The climax of Brunel's ambitions came with the SS *Great Eastern* at a length of 692 feet. It was his greatest achievement – and he reacted indignantly to any claim that it was not entirely his: 'I cannot allow it to be stated, apparently on authority, while I have the whole heavy responsibility of its success resting on my shoulders, that I am a mere passive approver of the project of another, which in fact originated solely with me, and has been worked out by me at great cost of labour and thought devoted to it now for not less than three years'.[37] Those photographs of Brunel, with the *Great Eastern's*

Isambard Kingdom Brunel in front of
the *Great Eastern*'s launching chains.
Photograph by Robert Howlett, 1857

launching chains in the background, cigar chomped firmly in his jaws and stovepipe hat on his head, sum up perfectly what the Victorian practical man of engineering was meant to be.

For most Victorians, this was simply what engineers were meant to look like. Men of invention were iconoclastic, obsessive, single-minded individuals – their invention their only goal. Men like these had the power to transform the world around them and create new tomorrows. That was the message of Samuel Smiles' *Self-Help* and his *Lives of the Engineers*. When new railway lines were opened, new bridges built and steamships launched, there was rarely any question about whose labour they were. The *Great Eastern* could not be mentioned without acknowledging Brunel. As successive attempts to get the 'Leviathan' into the water were made and failed, Brunel was the central character in the story. Less often acknowledged, as newspapers listed the dimensions of new ships, or the amount of earth shifted to lay the foundations of a new railway, was the back-breaking labour that went into these engineering feats. When writers of scientific romances sketched the character of their inventor protagonists, more often than not, they too shared these kinds of attributes. Men like Brunel and the Stephensons were the templates from which H.G. Wells or George Griffith drew their own fictional characters. Their stories revolved around inventors themselves and their obsessive relationships with their creations – and as scientific romances proliferated in books and popular magazines, so they helped feed their readers' understanding of what invention, and inventors, looked like.

Owning invention

In 1815, George Stephenson was still an obscure engine-wright at Killingworth Colliery. While he was working on building the *Blücher*, he was also tinkering away at another invention. Like all other deep mines, Killingworth was under constant threat from explosions as volatile coal gas came into contact with the naked flames from miners' lamps. After experimenting with different designs, Stephenson developed a lamp that could be used safely in mines without causing explosions. It was put on show at the Newcastle Literary and Philosophical Society on 5 December 1815. 'From eighty to a hundred of the most intelligent members of the Society,' were at that meeting, according to Samuel Smiles in his biography of Stephenson. Stephenson himself, 'proceeded, in his strong Northumbrian dialect, to describe the lamp. Down to its minutest details'. He performed experiments for them, with coal gas precariously collected from the Killingworth mine, showing how his lamp could be used safely without causing an explosion. The performance, and Stephenson's own earnest and impressive manner 'exciting in the minds of his auditors the liveliest interest both in the inventor and his invention', and the 'Geordy lamp' was soon in regular use in the local mines.[38]

At more or less the same time, Humphry Davy had developed his own safety lamp at the Royal Institution. Davy had been approached during the summer of 1815 by a group of mine owners and others to see whether he might be able to offer any solution to the problem of explosions in mines. 'There appears to me to be several modes of destroying the fire-damp without danger,' Davy told them. 'I have thought of two species of lights which have no power of inflaming the gas which is the cause of the fire-damp, but I have not here the means of ascertaining whether they will be sufficiently luminous to enable the workmen to carry on their business.' The eventual result of Davy's investigations was his paper 'On the Fire-damp of Coal-mines, and on Methods of Lighting the Mines so as to Prevent its Explosion', read before the Royal Society on 9 November 1815. Inevitably, Stephenson soon found himself accused of plagiarising Davy's invention. As Smiles later put it: 'What chance had the unknown workman of Killingworth with so distinguished a competitor? The one was as yet but a colliery engine-wright, scarce raised above the manual labour class, without chemical knowledge or literary culture, pursuing his experiments in obscurity, with a view only to usefulness; the other was the scientific prodigy of his day, the pet of the Royal Society, the favourite of princes, the most brilliant of lecturers, and the most popular of philosophers'[39]

This was a clash between competing cultures of knowledge. As far as Davy's Royal Society backers were concerned, it was simply inconceivable that a mere artisan could have achieved what Stephenson had done. That was not how invention worked. 'It will hereafter be scarcely believed,' spluttered John Ayrton Paris, Humphry Davy's biographer, 'that an invention so eminently philosophic, and which could never have been derived but from the sterling treasury of science, should have been claimed in behalf of an engine-wright of Killingworth, of the name of Stephenson – a person not even professing a knowledge of the elements of chemistry.'[40] Davy even had a notice, signed by 'the first chemists and natural philosophers of the country, with the President of the Royal Society, the most illustrious body in Europe, at their head', in the local press to deny Stephenson's claims to invention. 'It is disagreeable to be thus obliged to use artillery for the destruction of bats and owls,' he huffed, 'but it was necessary that something should be done.'[41] Ironically, just a few years later, Michael Faraday, then Davy's laboratory assistant, was accused of stealing the idea of electromagnetic rotation from William Hyde Wollaston (who was one of the signatories of the letter damning Stephenson). Just as it was inconceivable to Davy that an 'engine-wright' could do anything so original, it was inconceivable to Wollaston's supporters that a mere mechanic could come up with something so philosophical.[42]

Another similar example of competing claims to invention was the toxic debate that took place between the astronomer James South – a fellow reformer and close friend of Charles Babbage and John Herschel – and his instrument maker Edward Troughton. Troughton was widely respected and recognised as one of the best makers of telescopes in the business. Relatively unusually for an instrument maker, he was not only a fellow of the Royal Society but had been awarded the society's prestigious Copley Medal in 1809 for his work on astronomical instruments. South commissioned Troughton to build the mounting for a new telescope, but he insisted that the instrument Troughton had made for him was unfit for the purpose it was meant for and refused to pay for it. The problem boiled down to the fact that South, supported by other members of the Astronomical Society like Babbage and Herschel, simply refused to accept Troughton's expertise as a mere instrument maker. The clash ended up in court and ultimately led South to finally pay Troughton's bill. South expressed his contempt for Troughton in a final calculated insult, breaking up the instrument (which had cost him well over £1,000) and selling it for scrap by public auction. In 1839, he advertised a 'quantity of Mahogany, other Wood, and Iron, being the Polar Axis of the Great Equatorial Instrument made for the Kensington Observatory, by Messrs. Troughton and Simms'.[43] The rest was sold off a few years later, the posters for this sale fulminating about the 'mechanical incapacity of English astronomical instrument makers of the present day'.[44]

Clashes like these were relatively common throughout the first half of the nineteenth century. As much as anything else they were expressions of competing notions of where invention came from and who, properly speaking, owned it. Another good example is the partnership between William Fothergill Cooke and Charles Wheatstone that led to the first patent for an electromagnetic telegraph in 1837.

Cooke had been working independently on a plan for a commercially viable telegraph for some time before he approached Wheatstone and, after finding that he was engaged in a similar enterprise, suggesting a partnership. It was not long after the granting of a patent that the partners came to blows, however. The tensions between the two men revealed deep divisions (and insecurities on Cooke's part) around the business of invention and its crediting. Cooke was particularly incensed at seeing press reports ascribing the invention to Wheatstone alone and relegating himself to the role of business partner. He demanded that Wheatstone publicly repudiate the reports and acknowledge that it was Cooke who had the primary claim to be considered the instrument's inventor. All Wheatstone had done, according to Cooke, was fine-tune an instrument that was already to all intents and purposes a practical proposition. Unsurprisingly, Wheatstone demurred, insisting that without his expert intervention, the instrument would never have moved beyond the drawing board: Cooke simply lacked his knowledge of how electricity worked.[45]

The arbitrators' report adjudicating the respective claims to the invention of the two partners makes interesting reading. In an (eventually abortive) effort to resolve their differences, Cooke and Wheatstone called in Marc Isambard Brunel and the chemist John Frederic Daniell to apportion credit where credit was due. They were chosen to represent the two sides of the coin: the practical man joining with the gentlemanly professor from King's College London to ponder the nature of invention. The arbitrators' final statement was a masterpiece of diplomacy: 'While Mr Cooke is entitled to stand alone, as the gentleman to whom this country is indebted for having practically introduced and carried out the Electric Telegraph as a useful undertaking, promising to be a work of national importance; and Professor Wheatstone is acknowledged as the scientific man, whose profound and successful researches had already prepared the public to receive it as a project capable of practical application; it is to the united labours of two gentlemen so well qualified for mutual assistance, that we must attribute the rapid progress which this important invention has made during the five years since they have been associated.'[46] The arbitration succeeded in at least temporarily dampening Cooke's fury by drawing a carefully calculated line of demarcation between the stories Cooke and Wheatstone themselves told about who did what, and of what actually constituted invention. It meant that both parties had some sense of vindication. Wheatstone had done one thing, Cooke had done another – and it took both of them to make a successful telegraph.

Wheatstone was soon under attack from another direction for allegedly stealing another man's intellectual property. The Scottish clockmaker Alexander Bain accused him of having stolen his invention of the electromagnetic clock. Bain claimed in a series of letters to the *Literary Gazette* that he had approached Wheatstone with the invention to further his prospects in London, only to find Wheatstone patenting the idea. The *Mechanics' Magazine* enthusiastically took up the cudgel on Bain's behalf, portraying the dispute as one between the 'humble watchmaker from Caithness' and an unscrupulous gentleman of science.[47] The *Mechanics' Magazine* took an identical stance when they championed the cause of Liverpool gilder Thomas Spencer and his claim to have been the first to introduce the new technology of electrometallurgy to England. The journal complained 'that there still exists in certain scientific circles the same dogged reluctance, on which we before animadverted, to do justice to the humble "Carver and Gilder of Liverpool" – for no better reason that we can discover, but the sin of being humble.'[48] The problem was an elite scientific refusal to give proper credit to the working man: 'In one [book], which is a Birmingham production, intended, like most Birmingham wares, for the million, and written by a person, who, like the hero of his subject, boasts the license of no learned society to be useful to his fellow men, the art is honestly admitted to have *"proceeded from the hands of Mr Spencer,"* ... while, in the other, which is a London production, dedicated to the Consort of Her Majesty, and written by a Fellow of the Royal Society, the name of Mr Spencer is never once mentioned!'[49]

What these disputes and others like them had in common was that they were fundamentally about expertise and its ownership. Who really made things happen? Whom did the future belong to? From its establishment in 1823, the *Mechanics' Magazine* had positioned itself as the voice of the practical, mechanical man. The magazine was there to defend people like George Stephenson, or William Fothergill Cooke, or Alexander Bain, from the attempts to deprive them of credit for their work. But just who the *Mechanics' Magazine's* imagined constituency really was was never entirely clear either. In some ways, it seems to have depended on who their opponent was in any particular debate. There were competing views, not just about who should get the credit for invention, but about how invention should be rewarded. The natural philosopher, lawyer and reformer of the Royal Society William Robert Grove argued that anyone who patented their scientific discoveries or inventions had 'no right to look for fame; to those who sell the products of their brains, the public owes no debt'.[50] In other words, they were hucksters, not men of science. Scientific fame, on the other hand, was just what people like Bain, Cooke or Spencer argued they deserved. They saw no distinction between making knowledge and making money.

There were real fractures here. The clashes between George Stephenson and Humphry Davy revolved around the issue of whether a mere workman could invent and whether the miner's practical knowledge of conditions underground could compete with the chemist's expertise. Similarly, the clash between James South and Edward Troughton was an attempt by one of the new breed of disciplined specialists to put the practical know-how of a skilled instrument maker in its proper place. Troughton and his supporters disagreed about where that proper place was. As far as they were concerned, an instrument maker had claims to expertise that were just as valid as those of a specialist astronomer. It was clear to Charles Wheatstone that only his specialist knowledge of electricity could have made the telegraph work. To William Cooke, on the other hand, it seemed just as clear that the telegraph belonged to him. Fundamentally, these were debates about who was going to shape the future – and how. There were no clear answers to that question. Gentlemen of science, engineers and mechanics all jostled for credit. But it was clear to everyone that a future made through invention was coming. And by the middle of the century there were more and more places where that future in the making really did seem to be starting to take shape.

Chapter 3

Measure for Measure

*'The Queen is convinced that the President will
join with her in fervently hoping that the electric
cable which now connects Great Britain with
the United States will prove an additional link
between the nations, whose friendship is founded
upon their common interest and mutual esteem.'*

That was the message that Queen Victoria sent to James Buchanan,
President of the United States, on 16 August 1858, to inaugurate the newly completed Atlantic Cable. A little later, Buchanan
responded. 'It is a triumph more glorious,' he averred, 'because far
more useful to mankind, than was ever won by conqueror on the
field of battle.' There were more pious sentiments to follow: 'May
the Atlantic telegraph, under the blessing of Heaven, prove to be
bond of perpetual friendship between the kindred nations, and an
instrument destined by Divine Providence to diffuse religion, civilisation, liberty and law throughout the world.'[1] These fine words
were a reflection of just how clearly politicians on both sides of the

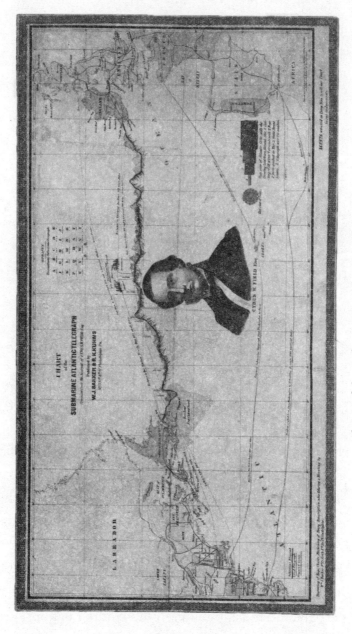

Chart of the Submarine Atlantic Telegraph.
W. J. Barker and R. K. Kuhns (Philadelphia, 1858)

Atlantic understood what a significant moment this was. It was a triumph of engineering and the practical application of the latest scientific knowledge. Its accomplishment had required organisation and resources on an unprecedented scale. The cable was a harbinger of a new kind of future – and it was all the more shocking, therefore, when a few weeks after those triumphant messages sped back and forth, the cable stopped transmitting.

Short-lived the triumph might have been, but the Atlantic Cable really was a huge enterprise. In the first place, it was an astonishingly ambitious project. After all, it was barely two decades since the telegraph had been invented. Even as telegraph engineers were still struggling with the realities of building cable networks overland, they were fantasising about something far bigger and more far-reaching in its potential. A telegraph cable across the Atlantic would be the first link in a chain of cables that would encircle the globe. It would revolutionise communication, it would make the world seem smaller and, so the optimists hoped, bring about a new era of international cooperation and understanding. It would be 'an enterprise which is destined to do more towards the realisation of a millennium of love among men than the efforts of all the diplomatists and missionaries are ever likely to accomplish', according to one New York newspaper.[2] Simply as an engineering project, its scale was tremendous. A way would need to be found to stretch a cable several thousand miles in length across the ocean floor between two continents. As things turned out, the failure of the first attempt had far-reaching consequences, too. As telegraph engineers and men of science struggled to understand the cable's failure, they started to imagine new ways of doing things and new ways of working that would prove crucial for turning their dreams of a technological future into reality.

The 1858 cable had been Cyrus Field's dream. Field was very

much – certainly in his own view of himself – a self-made man. The son of a Congregationalist minister in Stockbridge, Massachusetts, born in 1819, he had arrived in New York at the age of fifteen with only $8 in his pocket (about $250 in today's money). He was apprenticed as a clerk in a dry goods merchant's shop for three years before returning home to join his older brother's paper manufacturing business. By 1840, Field had set up his own paper manufacturing concern and then joined the New York wholesale paper merchants E. Root & Co. When that firm failed, Field set up on his own again, and by the 1850s had amassed enough of a fortune to retire and build a substantial mansion for himself in New York's fashionable Gramercy Park.[3] It was a classic American tale of a rise from (comparative) rags to riches through hard grind and determination – and a story that would have appealed to Samuel Smiles on the other side of the Atlantic, too. Field celebrated his retirement and his fortune with an expedition to South America in company with the landscape painter Frederic Edwin Church. He was following in the footsteps of his hero, the German naturalist and explorer, Alexander von Humboldt.[4] Arriving back in New York at the end of the expedition, Field had a live jaguar and 24 parrots in his baggage.

Shortly after Field's return from South America, he was introduced to the telegraph engineer Frederic Newton Gisborne, who was looking for backers for his plan to lay an underwater cable from Newfoundland to New York. It was the beginning of the dream that would become the Atlantic Cable. Born in Lancashire in 1824, Gisborne had arrived in Canada in 1845. He was soon engaged in telegraph speculation, helping to form the British North American Telegraph Association to build lines along the Canadian Atlantic seaboard. By the beginning of the 1850s, underwater cables were already being built in Europe, and it seemed clear to many North American telegraphic speculators that underwater telegraphy along the Atlantic coast offered plenty of scope for potential profit. Retired or not, Field was not a man to ignore such an opportunity, or the prospect of even greater profit if the dream of an Atlantic Cable could be realised. From Gisborne's point of view, Field was a celebrated and successful man of business – an embodiment of the American dream of self-improvement. He looked like a man who could get things done, and who knew other men who were just as experienced at the business of success.

Field duly recruited the industrialist Peter Cooper, his son-in-law Abram Stevens Hewitt and the wealthy banker Moses Taylor, along with the famous telegraph inventor Samuel Finley Breese Morse, to make up a self-styled 'Cable Cabinet' to turn the dream into a reality. These were men whose reputations rested on their ability to get things done. Between them, they wielded serious power and influence – just the kind of men who knew how to negotiate the corridors of commercial and political power. Field was soon acquiring allies on the other side of the Atlantic, too – British backers were essential to the project. Summer 1856 found Field in London, talking to the telegraph engineer Charles Tilston Bright, who, as the Magnetic Telegraph Company's engineer, had overseen laying telegraph cables beneath the Irish Sea, John Brett, whose Submarine Telegraph Cable had laid the first telegraph line linking Britain to the rest of Europe, and George Parker Bidder from the Electric Telegraph Company – which was originally founded by the telegraph's inventors, Charles Wheatstone and William Fothergill Cooke. Isambard Kingdom Brunel was another supporter of the project, as was Michael Faraday. This litany of big names simply emphasises just how much influence, money and power had to be mobilised to get the cable going. But by the end of the year, Field had founded the Atlantic Telegraph Company with £350,000 of capital and hired Edward Wildman Whitehouse as his chief electrician.

Hiring Wildman Whitehouse was, in many ways, a very idiosyncratic decision on Field's part.[5] Wildman Whitehouse had been trained originally as a surgeon and had a lucrative and successful medical practice in Brighton. But during the early 1850s he had started to carry out his own experiments in telegraphy – and he was particularly interested in the problems faced by underwater cables. His efforts had impressed John Brett enough that he lent him long lengths of cables for his experiments. When he presented the results of these investigations at the British Association for the Advancement of Science's annual meeting in Glasgow in 1855, he made it very clear that he was speaking to the gathering neither as a philosopher nor an engineer. He was very much the practical, no-nonsense man, who wanted a simple answer to a simple question: was underwater telegraphy over long distances, such as those separating Britain from its colonies, commercially viable? 'Can it be proved to be practicable – that is, commercially practicable – and capable of working at such a speed as will admit of messages being sent at a low tariff?' With that question foremost, he had 'investigated the phenomena exhibited by electrical currents in subterranean and submarine wires, as a speciality, and with a direct leaning towards their practical application, rather than in their more general and more extended theoretical aspects.'[6] Wildman Whitehouse's answer to the question was an emphatic yes: long-distance underwater telegraphy was perfectly possible. It was just the answer that Field wanted to hear.

On 7 August 1857, the USS *Niagara*, on loan from the US Navy, steamed out of Valentia Bay on the west coast of Ireland with Wildman Whitehouse and more than 1,000 tons of copper cable, wrapped in a sheath of iron wires and the insulating material gutta-percha, on board. She was accompanied by HMS *Agamemnon*, on loan from the British Navy, carrying its own heavy load of cable.

The plan was that at mid-Atlantic the cable would be spliced, and the *Agamemnon* complete the voyage to Newfoundland – neither ship was big enough to carry the entire cable on its own. At 4pm on 10 August, Whitehouse reported that nearly '360 miles have now been successfully laid down into the sea', and that the 'signals are everything an electrician could desire'. A few hours later, the cable broke. They tried again the following year, this time with the two ships starting their voyage in mid-Atlantic, splicing their cables and setting off towards their respective destinations. The splicing was carried out on 29 July, and the *Agamemnon* and the *Niagara* both arrived at their destinations – Valentia Bay in Ireland for the one, and Trinity Bay in Newfoundland for the other – on 5 August. Field, who had been on the *Niagara*, telegraphed laconically to his wife: 'Arrived here yesterday. All well. The Atlantic telegraph cable successfully laid.'[7]

It was an astonishing achievement – and completed against the odds. Some of Field's strongest early supporters had wanted to back out. 'Tomorrow the hearts of the civilised world will beat to a single pulse,' said the *Evening Post*, 'and from that time forth forevermore the continental divisions of the earth will now in a measure lose those conditions of time and distance which now mark their relations to the other.'[8] Just a few days later, Henry Ward Beecher marvelled 'how strange it will seem to have that silent cord lying in the sea.' It was going to change the world: 'Markets will go up and fortunes made down in the depth of the sea. The silent highway will carry it without noise to us. Fortunes will go down and bankruptcies spread dismay, and the silent road will bear this message without a jar and without disturbance.' He thought the prospect was 'in some sense, sublime.'[9] The news was greeted in New York by a 100-gun salute, and a parade with fireworks on 1 September. *Harper's Weekly* resorted to bad poetry:

Come listen all unto my song,
It is no silly fable;
'Tis all about the might cord
They call the Atlantic cable

...

And may we honor evermore
The manly, bold, and stable,
And tell our sons, to make them brave,
How Cyrus laid the cable.[10]

But by the time those lines were published on 11 September, things were already looking less than rosy for the Atlantic Cable's future. Even those congratulatory messages passed between queen and president were not quite what they appeared. They had only been transmitted legibly with a great deal of difficulty. In fact, by the beginning of September 1858 it was clear that the cable had failed completely, just as the company was hoping to open it up for public use. The cable might still span the ocean, but it was no longer performing its duty. By 6 September, the failure was public knowledge. A letter in *The Times* in London admitted that 'owing to some cause, at present not ascertained, but believed to arise from a fault in the cable at a point hitherto undiscovered, there have been no intelligible signals from Newfoundland since 1 o'clock on Friday morning the 3d inst'. A party of 'various scientific and practical electricians' were 'investigating the stoppage with a view to remedy the existing difficulty'.[11] That turned out to be a forlorn hope. It soon became clear that the cable was beyond repair. Fingers of blame were already pointing – and they were pointed at the company's chief electrician, Wildman Whitehouse.

The relationship between Wildman Whitehouse and the Atlantic Telegraph Company's directors had been fragile for some time. The qualities that had attracted Cyrus Field to him in the first place – his sense of himself as a proudly self-made man and an outsider in the worlds of engineering and electricity – now looked like vulnerabilities. There was nothing subtle about his views on electricity. They were robustly practical and common sense. He insisted on using his own patented instruments to transmit and receive through the cable and when the signals started fading his response was simply to increase the flow of electricity. He had clashed repeatedly with William Thomson, the professor of natural philosophy at the University of Glasgow, and one of the company's chief scientific advisers. Born in Belfast, raised in Glasgow, and rigorously drilled in mathematics at Cambridge, Thomson was a new kind of natural philosopher. His father, James Thomson, was professor of mathematics at Glasgow, and his brother (also James) apprenticed as an engineer. William Thomson possessed a rare blend of mathematical flair and talent as an experimenter. After completing his mathematical studies at Cambridge, he had spent time in Paris, working in Henri Victor Regnault's laboratory to hone his experimental skills, before returning to his native Glasgow as the university's new professor of natural philosophy. He was not a man inclined to suffer fools gladly.[12]

Coming as he did from industrial, Presbyterian Glasgow, Thomson was convinced that his science was meant to be useful. Putting science to work was a way to godliness. The careful study of physical processes was a way of revealing the ways in which nature worked most efficiently – and that knowledge could then be put to good use in making human industry less wasteful and more productive.[13] At Glasgow, Thomson forged links with hard-nosed industrial men who thought, as he did, that success lay in careful attention to detail. The key to making his science useful was careful, detailed and accurate experimentation, and it was that approach which convinced Thomson that submarine cables had properties entirely different to subterranean ones and needed to be treated differently. Rather than increasing the amount of electricity pumped through the cable in response to failing performance, it was better to use sensitive equipment, like his own mirror galvanometer, that could detect the feeble signals. Wildman Whitehouse ignored him or pretended to. When it became clear that, despite his bluster, it was only by using Thomson's apparatus rather than his own that Wildman Whitehouse had been able to keep the cable operational at all, his days as chief electrician were numbered.

It was not until 1865 that Field and the Atlantic Telegraph Company were in a position to try again. The outbreak of civil war in the United States made things more difficult, but the company and its supporters also needed to raise more capital. Even as the war was raging, Field and his backers were lobbying hard on both sides of the Atlantic and scrambling to persuade new investors. 'Some days I have worked from before eight in the morning until after ten at night to obtain subscriptions to the Atlantic Telegraph Company,' Field wrote in one letter.[14] Everything invested in the broken cable beneath the Atlantic was lost beyond recovery – and it was also becoming clear that one of the reasons for the failure was the quality of the cable itself. A new cable was needed – and this time more attention would need to be given to making sure that it was all made to the same standard. As plans were laid during the early months of 1865, it was decided that this time the entire cable would be carried on one ship, so Brunel's *Great Eastern* was chartered. Brunel's leviathan steamed away from Valentia on 23 July with 7,000 tons of cable on board. All seemed to be going well until 2 August, when disaster struck and the cable broke. Several attempts were made to recover it but, in the end, it had to be abandoned as well.

They tried again the following year – and this time their luck prevailed. They even managed to retrieve the lost cable abandoned the previous year. On 27 July 1866, Field wrote to his wife: 'All well. Thank God the cable has been successfully laid and is in perfect working order.'[15] *The Times'* correspondent announced: 'This evening, at about 5 o'clock, English time, the cable was completed between Europe and America.' And not just that: 'On Monday, at the latest, it is expected that the wire will be opened to the public, and the messages sent exactly in the order of priority in which they are received.'[16] This time, there would be no congratulatory messages between queens and presidents. The cable would be open for business. William Thomson duly gained a knighthood for his part in the triumph – 'I have no doubt he likes it', his fellow natural philosopher, John Tyndall, wrote rather cattily to the German man of science, Rudolf Clausius.[17] Just a few weeks later, William Robert Grove, in his presidential address to the British Association for the Advancement of Science, included the successful laying of the cable in his list of the year's greatest scientific achievements. He crowed to his audience that 'it is but a month from this time that the greatest triumph of force-conversion has been attained. The chemical action generated by a little salt-water on a few pieces of zinc will now enable us to converse with inhabitants of the opposite hemisphere of this planet'. 'The Atlantic Telegraph is an accomplished fact', he declared.[18]

Grove was deliberately underplaying to his audience. He knew perfectly well that far more than a bit of chemistry had been needed to make the cable 'an accomplished fact'. The entire project – both the repeated failures and the eventual success – had offered vital lessons in the business of making the future real. To make a success of the Atlantic Cable, its promoters had needed to organise and deploy resources on a massive scale. The navies of both Britain and the

United States had made ships available for the enterprise. In the end, they had needed the biggest ship that Victorian engineering ingenuity could provide – Brunel's *Great Eastern* – and the leviathan's role was as symbolic as it was practical in underlining the project's sheer bravura. They had needed to introduce new ways of manufacturing cable to ensure its quality. They had needed the resources of empire. But they had also needed to reflect hard on what kind of knowledge was needed to turn dream into reality – and who could be trusted to provide it. The ebullient, self-taught Wildman Whitehouse had failed to deliver. Much as the Victorians were wedded to the image of the individual, iconoclastic inventor – and it was an image that remained seductive throughout the century and beyond – the lesson learned from the Atlantic Cable was that what really worked was discipline. Behind the dogged individualism there needed to be a system. Genius on its own was not enough.

Natural standards

According to Robert Kalley Miller, fellow of Peterhouse, Cambridge, and later author of *The Romance of Astronomy*: 'The success of the Atlantic Cable is in great measure the result of years of patient work in the Glasgow Laboratory.' The laboratory Miller was talking about had been established by William Thomson at Glasgow University, shortly after he had been appointed professor of natural philosophy in 1846. Thomson established his laboratory to make sure that his students were taught the discipline of experiment. Rather than simply watching their professors do the work, as was the common practice at other universities, Thomson's students would gain experience by doing it for themselves. The work they learned to do needed systematic care and attention to detail. Throughout his long career at Glasgow, Thomson invented and patented scientific instruments – many of them, like his mirror galvanometer, designed with telegraphy in mind. He worked closely with Glasgow's instrument makers, making use of their resources and their expertise.[19] Miller had been one of Thomson's students, and even took his place teaching in the laboratory when the professor was ill. He recalled that he and his fellow students spent their time 'testing and perfecting the numerous exquisite instruments of Sir William Thomson's invention'. He thought that 'the excellent electrometers turned out by the Glasgow makers owe much of their value to the fact that each one has been carefully tested and regulated in the University laboratory before it is sent out for service'.[20] It was another example of Thomson's conviction that careful and disciplined measurement was vital for proper and useful knowledge. Without measurement, 'your knowledge is of a meagre and unsatisfactory kind: it may be the beginning of knowledge, but you have scarcely, in your thoughts, advanced to the stage of *science*, whatever the matter may be'.[21]

Thomson was not the only one to believe in the value of disciplined measurement. In 1861, Charles Bright and Latimer Clark – both electricians who had worked for Cyrus Field on the Atlantic Cable – delivered a paper at the annual meeting of the British Association for the Advancement of Science that was essentially a manifesto for standards. It argued for 'the desirability of the establishment of a set of standards of electrical measurement', and asked for 'the aid and authority of the British Association in introducing such standards into practical use'.[22] They were knocking at an open door. The BAAS leadership was well aware of the problems of quality control that were dogging the Atlantic Telegraph Company and the telegraph industry more generally. They promptly established a committee to look into the matter. Over the next few years, the committee delivered a series of reports as they investigated and experimented on the different possibilities. A reliable unit of electrical resistance was the obvious starting point – and it had to be robust and portable if it was going to be effective. By definition, units needed to be the same in every laboratory and every telegraph office and workshop. Thomson, of course, was a key member of the committee.

James Clerk Maxwell was another key member of the committee. Like Thomson, Maxwell was the product of rigorous mathematical training at Cambridge and had been professor of natural philosophy at King's College London since 1860. Another Scotsman, he shared many of Thomson's views about the importance of efficiency and utility in science.[23] Following Thomson's advice, he had devoted himself to electromagnetism, working his way through Michael Faraday's *Experimental Researches in Electricity* picking up on his strange ideas about forces and fields, and rewriting them in the language of mathematics. It was Maxwell at King's College who carried out much of the hard experimental work for the BAAS committee. He built specialist apparatus and the instrument-making firm Elliott Brothers turned them into reliable commercial instruments for telegraphy. Like Thomson, Maxwell knew that making the new association units useful needed training and discipline. Whenever 'many persons act together, it is necessary that they should have a common understanding of the measures to be employed', he insisted, along with telegraph engineer and fellow committee member Fleeming Jenkin.[24] The theoretical work on electromagnetism that Maxwell carried out while he was at King's went hand in hand with this practical work. Like Thomson, Maxwell moved easily between theory and experiment. He moved easily between university laboratory and electricians' workshop too. Useful and disciplined knowledge belonged in the one as much as the other.

In 1870, after much dithering, Cambridge decided that the university needed to establish a professorship in experimental physics. They recognised, too, that the 'founding of a Professorship would be incomplete unless means were also supplied to render the

Professor's teaching practical, and assistance given to him, both in the Laboratory and the Lecture-room'. That meant establishing a teaching laboratory like Thomson's at Glasgow – and Thomson was the first man they approached for the job. He turned them down and recommended Maxwell. Maxwell was dubious at first – he was independently wealthy and living on his Scottish estates at the time – but he was eventually persuaded. He worried that the 'class of Physical Investigations, which might be undertaken with the help of men of Cambridge education, and which would be creditable to the University, demand, in general, a considerable amount of dull labour which may or may not be attractive to the pupils'.[25] William Cavendish, Duke of Devonshire, and the university's chancellor, put up the money. Maxwell devoted meticulous attention to designing his new laboratory, visiting Thomson in Glasgow to see just what was wanted. The result was a building wholly designed for experiment. The Cavendish Laboratory was Maxwell's (and Thomson's) vision of the kind of disciplined knowledge that would make the future, cast in stone.

But there was more to Maxwell's vision of the place of standards in the making of the future. If that were all there was to it, then the 'Laboratory may perhaps become celebrated as a place of conscientious labour and consummate skill; but it will be out of place in the University, and ought rather to be classed with the other great workshops of our country, where equal ability is directed to more useful ends'.[26] Indeed, Maxwell worried that if his lab looked too much like an industrial factory, 'we may bring the whole university and all the parents about our ears'. The rich and powerful did not send their sons to Cambridge to turn them into industrial drudges – they sent them there to learn how to rule an Empire. Maxwell could reassure them that measurement was a higher calling than that – it was a way of understanding God. On Earth, measurement

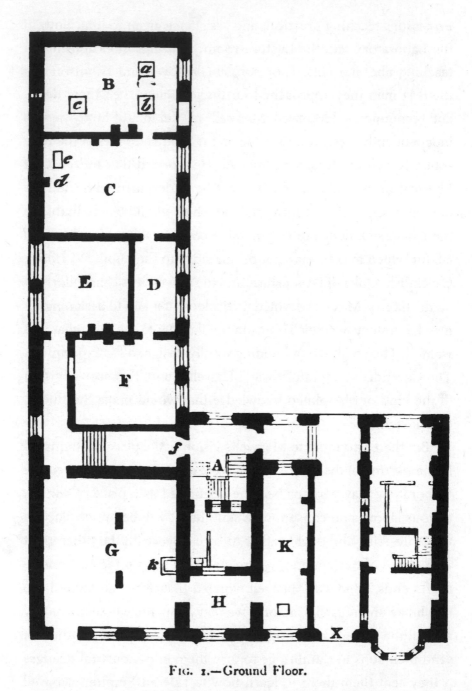

FIG. 1.—Ground Floor.

Ground floor plan of the Cavendish Laboratory.

Nature, 1874, 10

and standardisation were the best ways of ensuring the efficiency of industrial mass production, but they were also the tools the Creator had used to ensure the uniformity of nature's mass production. Godly men (such as those attending Cambridge as students) had 'aspirations after accuracy in measurement', simply because 'they are essential constituents of the image of Him who in the beginning created, not only the heaven and the earth, but the materials of which heaven and earth consist'.

As Maxwell took up his duties at the Cavendish Laboratory in 1874, he also saw publication of his *A Treatise on Electricity and Magnetism*. It was a volume (two volumes, in fact) that had been a long time in the making. Maxwell had been labouring at it almost since his student days – struggling to turn electricity into a mathematical science. Electricians like Wildman Whitehouse prided themselves on their hands-on practical knowledge. They knew how their instruments worked because they worked with them every day. The problems that had dogged the Atlantic Cable seemed to Maxwell, like Thomson, to show that this was unsatisfactory. This string and sealing wax approach needed to be replaced by measurement, disciplined training and systematic knowledge. The BAAS standard units supplied the first, and places like the Cavendish the second. Maxwell's *Treatise* was meant to complete the triangle. At its core was the ether – the mechanical medium that filled all space and through which all electromagnetic energy travelled in waves. What the *Treatise* did was describe the properties of the ether and explain how the whole range of electromagnetic phenomena could be understood in terms of those properties. It turned telegraphy into a science and brought – or tried to bring – system to chaos.

In his inaugural lecture at the Cavendish, Maxwell laid his programme out clearly. He drew a contrast between experiments of illustration that aimed 'to present some phenomenon to the senses of the student in such a way that he may associate with it some appropriate scientific idea' and experiments of research in which 'the ultimate object is to measure something which we have already seen – to obtain a numerical estimate of some magnitude'. It was the latter kind: 'those in which measurement of some kind is involved', that were 'the proper work of a Physical Laboratory'.[27] Reviewing his friend Fleeming Jenkin's latest textbook a year previously, Maxwell had endorsed wholeheartedly the way in which the 'author of this text-book tells us with great truth that at the present time there are two sciences of electricity – one that of the lecture-room and the popular treatise; the other that of the testing-office and the engineer's specification. The first deals with sparks and shocks which are seen and felt, the other with currents and resistances to be measured and calculated.' That second kind of electricity added up to 'a sort of floating science known more or less perfectly to practical electricians'.[28] What Maxwell wanted to do was to give that 'floating science' some discipline. *A Treatise on Electricity and Magnetism* was an attempt to impose a mathematician's order on the messy world of electrical engineering.

Discipline and knowledge

By 1879, Maxwell was dead – of cancer at the early age of 48. His successor at the Cavendish, Lord Rayleigh, entirely shared his conviction that discipline was the way to make knowledge useful. John William Strutt, third Baron Rayleigh, had been one of Maxwell's strongest advocates when the search for a Cavendish professor was on. Like both William Thomson and Maxwell, he had studied mathematics at Cambridge – in Strutt's case graduating top of his class as senior wrangler (Maxwell and Thomson had both been second wranglers) and winning the prestigious Smith's Prize as well. He was promptly elected a fellow of Trinity College.[29] Like both Thomson and Maxwell, too, he combined mathematical flair with a passion for experiment. In pursuit of that passion, he built and stocked his own laboratory at the family seat, Terling Place, in Essex. Independently wealthy, he could afford to pursue precision experiment on his own account. He dabbled in acoustics and in optics, combining careful experiment with mathematical theorising. In 1871, for example, he came up with an explanation for why the colour of the sky is blue. He was interested in photography and carried out experiments in trying to produce diffraction gratings photographically. It was careful, precision work that needed delicacy and attention to detail.[30]

Following Maxwell's death, Thomson persuaded the baron that he was the best man to continue the tradition of disciplined experiment that Maxwell had inaugurated. If Rayleigh 'could see your way to take the Chair it would I am sure be much to the benefit of the university, and of science too', Thomson told him. Rayleigh left his laboratory at Terling with some reluctance – but there was an agricultural depression in progress and he needed the money. At the Cavendish, Rayleigh continued and improved on the Maxwellian tradition. He increased the number of undergraduates being trained in the laboratory and made their training more systematic. The laboratory was organised to take students through a regime that was as well calibrated to produce disciplined experimenters as the laboratory instruments themselves were to produce accurate measurements. Under Rayleigh's direction, 'each experiment was set out permanently on a table to itself, and written directions were provided. The classes were at regular hours, and a demonstrator was in attendance, who assigned the experiment, and gave help in any difficulty, finally approving or disapproving the numerical result.'[31] Some of the cadre of demonstrators would go on to run laboratories in their own right. Richard Glazebrook, for example, went on to be the National Physical Laboratory's first director.

Rayleigh's Cavendish was a place for collaborative and disciplined research as well as training. As one of his star students, Arthur Schuster, recalled, Rayleigh's overriding aim was 'to identify the laboratory with some research planned on an extensive scale so that a common interest might unite a number of men sharing in the work'.[32] The project that united them all was to be a continuation of Maxwell's research on electrical standards, using the apparatus that had been built originally for the BAAS committee. It was the means to produce not just electrical standards, but men of science trained to a common standard too. And the Cavendish cohorts were, by and large, men. J.J. Thomson, Rayleigh's successor as professor, described the laboratory as a 'place to which men who had taken the Mathematical Tripos could come, and after a short training in making accurate measurements, begin a piece of original research'. Thomson had nothing good to say about the few women who dared to breach the Cavendish's hallowed walls (unlike Rayleigh, who asked Eleanor Sidgwick to help him with his experiments on resistance). Under his direction, the laboratory was expanded even further. A new wing was opened in 1896 with 'a very large room used for elementary classes in practical physics, for examinations in practical physics for the Natural Science Tripos and for entrance scholarships to the Colleges. Besides this there was a new lecture-room, cellars for experiments requiring a constant temperature, and a private room for the Professor'.[33]

It was not just in Britain that scientific training and disciplined research were being touted as answers to the problem of how to remake the future. Across Europe and beyond, this was the new road map. In Berlin, capital of the new German Reich after the Prussian trouncing of France in 1871, money was poured into physics – to the envy and concern of others. John Tyndall at London's Royal Institution fretted that 'you will find in the Berlin laboratory

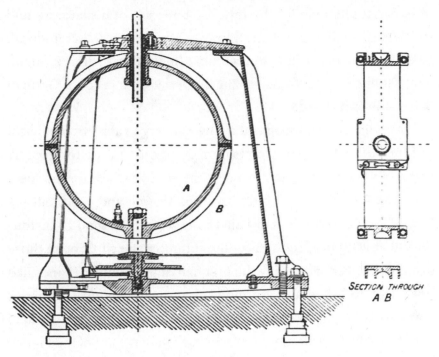

Lord Rayleigh's spinning coil apparatus for measuring the ohm.
Rayleigh's Scientific Papers (Cambridge: Cambridge University Press, 1900), vol. 2

the very things which my American and British friends and I should like to see in operation in all college and university laboratories in America and the British Empire.'[34] There was more to come. In 1884, the electrical industrialist Werner von Siemens announced his willingness to put his hand in his pocket to fund the establishment of an institution devoted entirely to research. 'England, France, and America, those countries which are our most dangerous enemies in the struggle for survival have recognised the great meaning of scientific superiority for material interests and have zealously striven to improve natural scientific education through pedagogical improvements and to create institutions that promote scientific institutions,' he claimed.[35] He could hardly have been more explicit about the central place physics was coming to occupy in imperial planning for the future. The result of his largesse and much hard lobbying was the *Physikalisch-Technische Reichsanstalt* (the Imperial Physical and Technical Institution) with Hermann von Helmholtz – celebrated as the 'Imperial Chancellor of German Science' – at its head.

Many in France thought that the country's catastrophic defeat in 1871 was at least partly a result of a national failure to properly embrace a scientific future. Bringing the world to Paris for the spectacle of international exhibitionism was all well and good, but the Second French Empire had failed to invest in the kind of institutions that were needed to make the future imagined through those exhibitions a reality. During the first half of the century, France had been held up by some in Britain as the model to emulate. Charles Babbage had argued in *Reflections on the Decline of Science in England* that the Royal Society should reform itself in the image of the French Academy of Sciences. Now though, those once admirable institutions looked moribund and out of shape for the times. In the wake of the Prussian defeat, it seemed clear that new institutions and more rigorous training regimes would be needed to restore French pride in their science as well as its uses in forging a new future for the new republic. Republican politicians like Jules Simon pushed hard for funding to improve the state of French science. By the end of the century, the chemist and politician Marcellin Berthelot reckoned that the 'life of today's savant is a many-sided one'. He was 'urgently solicited in the name of the public interest, in the most diverse spheres: specific applications in industry or national defense, public education, or even general politics'.[36]

By the second half of the century, increasing numbers of Americans were flocking to European universities and laboratories, eager for the kind of disciplined training they felt was lacking in their own institutions. As early as the 1850s, Charles Bristed was bragging to his fellow Americans about the sheer hard work and discipline required to study mathematics at Cambridge. 'For in the eight hours a day which form the ordinary amount of a reading man's study, he gets through as much work as a German does in twelve,' he said, 'and nothing our students go through can compare with the fatigue of a Cambridge examination.'[37] Men like Henry Rowland travelled to Berlin and the Cavendish to imbibe the disciplined approach to experiment that they could not get at home. 'Accurate measurement is an English science,' Rowland espoused. He returned from his European pilgrimage to make it into an American science too. Before his European trip he had already been appointed to the newly established professorship of physics at Johns Hopkins University. There he developed his laboratory along the lines of those he had seen in Europe. He developed precision instruments that made even British and German experimenters jealous. The 'Germans spread their palms, looked as if they wished they had ventral fins and tails to express their sentiments', when told about the accuracy of his apparatus.[38]

The British cult of disciplined laboratory work spread from the imperial centre to the peripheries of empire. As university physics departments and teaching laboratories sprang up, they were populated by the products of the Cambridge system. James Loudon, professor of physics at the University of Toronto, might not himself have been a graduate of the Cavendish, but he made sure that the systematic teaching of laboratory experimentation became an integral part of the curriculum. His protégé John McClelland was sent to work at the Cavendish with J.J. Thomson for a year before returning to set up a graduate laboratory. Henry Martyn Andrew, the first professor of physics at the University of Melbourne, had been 27th wrangler in the mathematics tripos in 1872. William Henry Bragg, appointed professor of experimental physics at the University of Adelaide in

1885, had been third wrangler in the Cambridge mathematics tripos in 1884. Even beyond imperial borders, the institutions of disciplined science looked like essential building blocks for the making of the modern state. Establishing an Imperial College of Engineering in Tokyo – headed by William Ayrton, a product of Thomson's Glasgow laboratory – was a key element in the newly restored Meiji regime in Japan's plans as they set about transforming the country from feudal state into an industrial powerhouse.

At the Paris International Electrical Exhibition in 1881, Britain and Germany went head-to-head over the matter of electrical standards. This was the first meeting of the International Congress of Electricians, and their chief aim at this inaugural gathering was to establish a set of internationally recognised standard electrical units. In the German corner were Werner von Siemens, whose deep pockets would soon be bearing the costs of establishing the *Physikalisch-Technische Reichsanstalt*, and Hermann von Helmholtz, who would be its first director. Representing Britain was William Thomson. The Germans pushed for international recognition of the standard unit of resistance that Siemens himself had developed, based on the length of a column of mercury. The British candidate, of course, was the BAAS unit as further refined by Lord Rayleigh and Eleanor Sidgwick at the Cavendish. Rayleigh had made the establishment of secure electrical standards the Cavendish Laboratory's key business when he took over. It was an exemplar of disciplined experiment. The apparatus had been set up on the ground floor to ensure stability. Measurements were taken late at night to avoid magnetic and other potential disturbances. The resulting unit of resistance embodied British notions of disciplined physics. 'It is an extraordinary and gratifying result for all of us,' wrote James Joule, one of the early pioneers of the theory of the conservation of energy, to Rayleigh.[39]

The battle over the ohm, as the unit of electrical resistance came to be called, simply emphasised how important disciplined science was seen to be by the final decades of the nineteenth century. Being able to claim ownership of standards was a matter of national prestige. Unsurprisingly, tempers frayed. At the 1881 meeting in Paris, the 'debate grew warm'. Non-aligned delegates were entertained by 'the unforgettable scene of comedy of Thomson and Helmholtz disputing hotly in French, which each pronounced *more suo*, to the edification of the representatives of other nationalities'. An inconclusive truce was called that year, but hostilities recommenced in 1882, with Thomson begging Rayleigh to attend too: 'We could get on but badly without you,' he worried. The Cavendish ohm was gradually winning the war of attrition though. It took a decade of bickering to achieve, but by 1891, the British could declare victory in the war of the standards. It was an important acknowledgement of the superiority of British scientific and industrial discipline in the race to the future. At the BAAS meeting of 1891, Oliver Lodge in his presidential address called for the establishment of a national physical laboratory to help maintain that edge over the competition. The committee established to investigate the matter was chaired by Lord Rayleigh. In due course, he made sure that it was his Cavendish protégé Richard Glazebrook who was appointed its first director to continue the business of maintaining standards and discipline alike in physics for the Empire.

In 1881, just as the opening skirmishes of the war of the standards were taking place in Paris, the British government established the Royal Commission on Technical Instruction 'to inquire into the instruction of the industrial classes of certain foreign countries in technical and other subjects for the purpose of comparison with that of the corresponding classes in this country; and into the influence of such instruction on manufacturing and other industries at home and abroad'.[40] Chaired by the industrialist and politician Bernhard Samuelson, the committee was a recognition that the future was too important to leave to chance. The committee members were nothing if not punctilious in their enquiries. Members toured Europe collecting evidence, amassing an impressive collection of pamphlets, school and college syllabuses and local, national and international exhibition catalogues. Their reports, published in 1882 and 1884, were voluminous (the second report filled five volumes) and comprehensive. They gained Samuelson a baronetcy and laid out the need to extend a disciplined scientific education through all echelons of British education, from universities down through technical colleges to schools. The future being drawn in scientific romances demanded it. Ownership of the future depended on having a citizenry that was at all levels thoroughly versed in science and its disciplines.

James Thomson's career – William Thomson's brother – offers a nice example of how things were changing. Just as interested in natural philosophy as his brother, after finishing his degree at Glasgow, James entered into an apprenticeship with the engineer and shipbuilder William Fairbairn. He set up business as a civil engineer in Belfast before being appointed professor of civil engineering at Queen's University Belfast. He stayed there until he returned to Glasgow as professor of civil engineering. His move from apprenticeship, to consulting engineer, to university chair was emblematic of the way engineering was being transformed from a craft to be entered through the traditional door of an apprenticeship, to a profession that demanded university training.[41] The cult of disciplined laboratory work that originated in Victorian physics laboratories was spreading by the end of the nineteenth century. Engineering now also needed the same kind of laboratory training. Chemistry laboratories, and physiology laboratories, were organised along the same lines. When Michael Foster was appointed by Trinity College, Cambridge to teach physiology in 1870, the physiology laboratory he worked to establish would adopt the kind of disciplined experimental regime that Maxwell and Rayleigh had put in place at the Cavendish.[42]

Disciplinary powers

Phileas Fogg, the main protagonist of Jules Verne's *Around the World in Eighty Days*, was the epitome of disciplined precision. The ultimate clockwork man, he 'breakfasted and dined at the club, at hours mathematically fixed, in the same room, at the same table, never taking his meals with other members, much less bringing a guest with him; and went home at exactly midnight, only to retire at once to bed. He never used the cosy chambers which the Reform provides for its favoured members. He passed ten hours out of the twenty-four in Savile Row, either in sleeping or making his toilet. When he chose to take a walk it was with a regular step in the entrance hall with its mosaic flooring, or in the circular gallery with its dome supported by twenty red porphyry Ionic columns, and illumined by blue painted windows.' As his new French servant observed: 'Mr Fogg seemed a perfect type of that English composure which Angelica Kauffmann has so skilfully represented on canvas. Seen in the various phases of his daily life, he gave the idea of being perfectly well-balanced, as exactly regulated as a Leroy chronometer. Phileas Fogg was, indeed, exactitude personified, and this was betrayed even in the expression of his very hands and feet; for in men, as well as in animals, the limbs themselves are expressive of the passions.'[43]

This was satire, a Frenchman poking fun at the passionless English. There was no satire to be found, however, in *The Soldier's Pocket-Book*, written by Major-General Sir Garnet J. Wolseley and first published in 1869. In it, the model for Gilbert and Sullivan's 'very model of a modern major-general' set out precisely how an officer and a gentleman should act. They should turn themselves into measuring machines, Wolseley suggested. Officers 'should endeavour to carry in their heads certain easy mathematical formulae regarding the measurement of distances, &c.'. They should make it a habit to 'note carefully, even as you whizz along in a railway carriage, the peculiar features of the country, the nature of its fences, &c.'. They should accustom themselves 'to time the pace at which you travel, to count the number of telegraph poles there are to a mile, and so ascertain how many yards they are apart, &c.'. Their bodies should be their instruments. Every officer should 'know the exact length of his foot, hand, cubit, and arms from tips of fingers of left hand, to right ear; he should know the height of his knee, waist, and eye, and also the exact proportion that his drinking-cup bears to a pint'.[44] Precision and accuracy was meant to be built into everything they did.

Indeed, Wolseley's strictures chimed very well indeed with Victorian notions of masculinity. It was meant to be the capacity for self-discipline that made the Victorian man. Arthur Conan Doyle made sure that these were the very sorts of attributes that characterised Sherlock Holmes. He was a man made of precision, and his popularity underlines just how ingrained the disciplinary ideal was by the closing decades of the century. Conan Doyle depended on his readers' instinctive recognition of what Holmes represented. For the stories to work they needed to know where he belonged and what he was. Holmes would have had no difficulty with the modern major-general's advice on the counting of telegraph poles. 'Holmes is a little too scientific for my tastes – it approaches to cold-bloodedness,' a friend warned Watson before he met his future friend for the first time. 'I could imagine him giving a friend a little pinch of the latest vegetable alkaloid, not out of malevolence, you understand, but simply out of a spirit of inquiry in order to have an accurate idea of the effects.'[45] Holmes was a disciplined mind in a disciplined body – and just like the Cantabrigian mathematicians who migrated to

Sherlock Holmes.
Sidney Paget, *The Strand Magazine*, December 1892

the Cavendish to learn how to discipline nature, he recognised self-discipline as a higher calling.

The lesson to be learned from the Atlantic Cable's failure and success, as far as many Victorians were concerned, was that these kinds of huge imperial enterprises depended on the combination of individual self-discipline and mass mobilisation. These were the new tools of empire, and the drive to remake knowledge and its institutions was very much part of this. Even before the Atlantic Cable, large-scale enterprises like the so-called magnetic crusades were being promoted as examples of the virtues of disciplined knowledge. Spearheaded by gentlemen of science and scientifically minded

Admiralty men, this was an ambitious enterprise to map systematically variations in the Earth's magnetism.[46] The goal was to increase knowledge and the reach of naval power at the same time. Just like the Atlantic Cable, the magnetic crusades depended on the deployment of accurate instrumentation of unprecedented precision. Naval expeditions sent out to map coastlines and chart the ocean floor required similar resources and served the same ends. The very possibility of laying a telegraph cable across the Atlantic depended on the fact that a systematic survey of the ocean floor had only recently been completed. Without that kind of knowledge of the submarine terrain, it would have been impossible to judge the best route and the length of cable required.

Maps on land depended on the systematic deployment of surveyors equipped with accurate apparatus as well. Although the Ordnance Survey had its origins in late eighteenth-century efforts to prepare for possible invasion by the French, by the middle of the nineteenth century it had become a substantial enterprise, dedicated to mapping the country inch by inch. Mapping was an imperial exercise in every sense. As early as 1819, there were calls for the Ordnance Survey to take on the mapping of Ireland, and work duly began in 1824. The East India Company established the Great Trigonometrical Survey in 1802 to map existing and prospective territories. Mapping was central to the business of controlling new ground – the survey was started just a couple of years after the company's armies finally defeated Tipu Sultan, the Tiger of Mysore.[47] Right at the end of the century, Rudyard Kipling's *Kim* offered a tantalising glimpse of the difficulties surveyors might encounter. Sent clandestinely to survey hostile territory, 'it was not an amusing trip from Kim's point of view, because – in defiance of the contract – the Colonel ordered him to make a map of that wild, walled city; and since Mohammedan horse-boys and pipe-tenders are not expected to drag Survey chains round the capital of an independent native state, Kim was forced to pace all his distances by means of a bead rosary.[48]

Kim's travails underlined the imperial purpose of maps and expeditions, and the scientific exhibitions sent out from London to various parts of the Empire and beyond at regular intervals throughout the century carried a similar message. They were exercises in discipline. In 1871, for example, coordinated observations of the solar eclipse were carried out at locations across India. As the expedition's leader, Norman Lockyer 'had arranged that observations should be made with instruments of the same nature precisely in Ceylon, India, Java, and Australia, so that the results might be strictly compared.'[49] These were spectacles of discipline in action as much as astronomical exercises: 'Astonished natives gathered around the tower, curious to learn what their European masters were doing with the big telescopes pointed at the sky.'[50] The power of predicting eclipses often featured in adventure stories, like H. Rider Haggard's *King Solomon's Mines*, as evidence of European superiority. Similarly, the 1874 transit of Venus expeditions were the products of several years' careful planning and disciplined preparation. Observers trained their eyes and minds to make sure observations were consistent (to deal with the personal equation, Victorians would have said). Five identical observing stations were built at the Royal Observatory in Greenwich and stocked with identical precision apparatus before being dismantled and shipped out to be reassembled on site at the expedition locations in Egypt, Hawaii, New Zealand and a couple of Pacific islands. Like the observers, the instruments were rigorously tested and standardised. The transit clocks, built by Messrs. E. Dent & Co., were 'tested for a long time at the Royal Observatory, Greenwich, under very trying variations of temperature.'[51]

Those 'astonished natives' would be submitted to a regime of rigorous and disciplined measurement as well. In his *The Races of Britain,* published in 1862, John Beddoe had produced a racial hierarchy of the British Isles' inhabitants, based on their physical characteristics. At the bottom of the pile, with prognathous jaws and primitive features reminiscent of earlier inhabitants, were the Welsh and the Irish. The English and Lowland Scots were examples of a more advanced breed of men. The Celtic races were 'negroid' according to Beddoe, and this kind of racial typology was soon exported to the Empire as a way of classifying and recording native populations.[52] Anthropometrists developed their own precision apparatus and techniques of measurement. They drew charts of hair and eye colour, measured skulls and cranial angles and the length of arms and legs. The Anthropological Institute – formed in 1871 from the rival Anthropological and Ethnological Societies – regarded itself and its members, with their classifying zeal, as servants of empire. By measuring its inhabitants, putting them in their proper place, just as Cavendish men might measure electrical resistances or the wavelengths of light from distant stars, they were displaying the discipline they thought made the Empire great, and imposing discipline on its peoples.

Not just the races of man, but the moral quality of men could be made visible through the lens of accurate and disciplined measurement. Criminologists were confident that they could classify criminal types as easily as they could racial ones. Physical characteristics like the width of the forehead or the distance between the eyes were potential markers of moral character. Police forces assiduously collected photographs of the criminal classes in efforts to classify them. The promoter of eugenics Francis Galton was an advocate of this approach. He thought that the capacity to identify the morally degenerate through their appearance offered a way of improving racial stock. He was a forceful advocate of the use of fingerprinting to identify criminals too – a technique originally pioneered in the

Raj and championed by Galton, who started his campaign with a lecture on personal identification at the Royal Institution in 1888.[53] When Conan Doyle's readers picked up their copies of *The Strand Magazine* to devour Sherlock Holmes' latest exploits, this was what made what they read familiar to them. They would recognise Holmes as one of these disciplined men, devoted to accuracy and precision. His obsessive classifying of cigar ash, and his insistence that an individual's past could be read from their appearance, marked him out as a member of the disciplinary club.

By the end of the nineteenth century, disciplined knowledge was at the heart of empire, and more than that, it reminded the Victorians of their place in the cosmos. Precise and accurate measurement revealed the grandeur – and the uniformity – of the Universe. The disciplined, laborious, meticulous work that was done in places like the Cavendish, and in workshops manufacturing uniformly engineered cables and coils, actually mirrored the workings of the divine mind. To those not at home in James Clerk Maxwell's universe, it could all look a little peculiar. The French philosopher Pierre Duhem recoiled in horror when he read Oliver Lodge's account of Maxwell's ether: 'In it there are nothing but strings which move around pulleys, which roll around drums, which go through pearl beads, which carry weights ... We thought we were entering the tranquil and neatly ordered abode of reason, but we find ourselves in a factory.'[54] But as far as Maxwell and his followers were concerned, that was exactly what vindicated the measured discipline of their physics. What he did in his laboratory, God had done in his. And just as the Universe was governed through discipline, so the Victorians governed their own affairs. The Atlantic Cable's triumph, and the successes that followed, showed just how this kind of organised and applied discipline could transform the future. It was a model of how to get things done in society and in nature.

It was a far cry from the leisured natural philosophy of a century ago. Not that Joseph Banks and his cronies had not thought their science should be useful, rather that they imagined that utility in a very different way. The early nineteenth-century generation of heroic engineering men had prided themselves on the single-minded perseverance that underpinned their triumphs, but their late Victorian descendants channelled their discipline in a more focused and collective way. The image of the expert that emerged by the end of the century was of a disciplined, measured man – and this was a wholly masculine culture of knowledge and its making. They were men whose minds and bodies had been honed for power. Whether they studied the ether, built railroads or administered colonial outposts, they shared the same collective view of how expert knowledge was meant to be applied. Sitting – and sometimes sitting uneasily – alongside that collective vision of how to get things done were visions of the future that the technological transformations wrought during the nineteenth century would ultimately deliver. Underpinning both the wielders of expertise and the fantasies of things to come were the resources of an empire. And there were an increasing number of places where people really could go to see that collective spectacle of tomorrow in the making.

Chapter 4

Showing Off

On 4 June 1832, a new exhibition opened its doors to the London public. The National Gallery of Practical Science, Blending Instruction with Amusement was on Lowther Arcade – a covered shopping arcade running from the Strand at one end to Adelaide Street at the other. It was an excellent spot for an exhibition, part of the redevelopment just completed by the architect John Nash that included nearby Trafalgar Square. It was one of the places where fashionable Londoners went looking for entertainment. The arcade's shops were full of 'costly and elegant bijoutrie exhibited for sale', just like 'those celebrated in the Arabian Nights, and other works of eastern fiction'.[1] Just across the arcade from the gallery was the philosophical instrument maker Edward Marmaduke Clarke's self-styled 'Laboratory of Science'. The gallery itself promised to 'receive, for public exhibition, subject to immediate return on demand, and meanwhile protected by every precautionary arrangement, Works of Practical Science, free from any charge', and to afford 'every possible encouragement and facility for the practical demonstration of discoveries in Natural Philosophy, and for the exhibition of new applications of known principles to mechanical contrivances of general utility'.[2] Anybody paying their shilling at the door, which was the usual price of entry to exhibition halls and similar shows, could not have failed to notice, though, that most of the inventions there belonged to a single inventor – Jacob Perkins. Not everyone was impressed: the 'notion of erecting, at an expense of many thousands of pounds, a hall to exhibit two or three inventions of a particular individual – not all original and some of them mere abortions – was, to say the least, exceedingly preposterous', huffed the *Mechanics' Magazine*.[3]

Perkins was an American. He had arrived in London from Philadelphia early in July 1819. Originally from Massachusetts and apprenticed to a goldsmith, by the time he was in his twenties, Perkins had started making a name for himself as an inventor of useful machinery. By 1795, he had been awarded a patent for improved nail-making machinery and established himself as a manufacturer. He was a highly skilled engraver as well, and later developed a lucrative business in printing. By 1816, he had set up a printing shop in Philadelphia and was printing banknotes. Following from his engraving work, Perkins had been working on designing a machine that would be capable of producing forgery-proof banknotes. The Bank of England had established a prize for just such an invention – and Perkins had come to London to win it. In London, Perkins distributed examples of his banknotes to potential backers. In his path, however, stood the formidably despotic figure of Sir Joseph Banks, the Royal Society's president. As we've seen, little got done in matters of scientific innovation like this without Banks' approval and patronage – and despite his best efforts, Perkins failed to get the Banksian seal of approval. Banks made it quite clear that, in his view, the winner of such a prestigious and lucrative prize could only be an English-born gentleman.

Undaunted, Perkins stayed in London. The machine for print-ing banknotes was not the only invention he had to sell. While in Philadelphia, Perkins had also been tinkering with steam engines. In particular, he had been trying to develop steam engines that worked at very high pressures. The main outcome of his efforts in that direction was the steam gun, patented in 1824. It was 'reserved for Mr Perkins to astonish us, even at a period when we imagined that the ingenuity of man had exerted itself to the utmost', said *The London Mechanics' Register* of 'Mr Perkins' Extraordinary Steam Gun'.[4] 'If we did not feel that the adoption of this mode of warfare would end in producing universal peace, we should be far from rec-ommending it', they said.[5] The invention even inspired poetry (of a sort):

> Five hundred balls, per minute, shot,
> Our foes in fight must kick the beam;
> Let Perkins only boil the pot,
> And he'll destroy them all by Steam.[6]

On 6 December 1825, Perkins organised a grand demonstration of his invention. 'The noise of each explosion was like that of a musket', recalled one onlooker, 'and when the discharges were rapid, there was a ripping uproar, quite shocking to tender nerves.' The Duke of Wellington was there – and the King's brother, the Duke of Sussex (soon to be president of the Royal Society). 'Wonderful, wonder-ful – d----d wonderful; wonderful, wonderful – d----d wonderful!' was the duke's verdict, but Perkins failed to persuade anybody to buy one of his machines.[7]

Establishing the National Gallery of Practical Science was a fur-ther attempt on Perkins' part to find a market for his inventions. Putting invention on show was not an entirely new idea. After all, just as Perkins was trying to persuade the Duke of Wellington of the steam gun's virtues, the Brunels were conducting lines of distinguished visitors beneath the Thames to see the Shield in action. Engineering exploits like this were meant for spectacle, and the steam gun was no exception. Perkins had his own experiences from his time in Philadelphia to draw on as he plotted his new exhibition. He would have been familiar with the popular Peale's Museum, established by the artist Charles Willson Peale in 1784. Peale's Museum had moved from being an exhibition space for his own artistic productions to displaying curious specimens of North American natural history and technological curiosities such as the physiognotrace – a device that could be used to semi-automatically produce silhouette portraits.[8] The version on show at Peale's Museum had been patented by John Isaac Hawkins who was in partnership with Peale to market the invention. Peale's Museum had played a key role in bringing technological innovation within the horizon of polite Philadelphian society and it is easy to imagine that Perkins was attempting to use the same approach at the National Gallery of Practical Science to attract the attention of London's moneyed elites.

There were London examples to follow as well, of course. In his autobiography, Charles Babbage recalled his fascination with the Silver Lady automaton that he was taken to see at the age of eight at John Merlin's Mechanical Museum, of which more later.[9] Wolfgang von Kempelen's mechanical, chess-playing Turk had been another popular exhibit. Fashionable London flocked to shows like the panorama or the phantasmagoria – exhibitions deliberately designed as technological curiosities to challenge the senses with spectacle. The panorama deployed carefully calculated architecture and optics to make artificial scenes look natural. Robert Barker, the panorama's inventor, had a specially designed building put up in Leicester Square, not far from where the Adelaide Gallery would be located, to show off the illusion. To be properly displayed, the panorama required 'a circular building or framing erected, on which this drawing or painting may be performed; or the same may be done on canvas, or other materials, and fixed or suspended on the same building or framing.'[10] The showroom had to be 'lighted entirely from the top, either by a glazed dome or otherwise, as the artist may think proper'; and the entrance to the viewing room had to be 'from below, a proper building or framing being erected for that purpose, so that no door or other interruption may disturb the circle on which the view is to be represented.'[11]

The phantasmagoria used optical illusions and hidden magic lanterns to put ghosts on stage and needed the same kind of careful management of space to work. Paul de Philipsthal, the phantasmagoria's promoter, offered the audience at the Lyceum Theatre, just off the Strand, a 'grand Cabinet of Optical and Mechanical Curiosities ... an entirely Novel EXHIBITION of OPTICAL and MAGICAL ILLUSIONS'. The show would have 'the merit of unmasking artful Impostors, and opening the eye of such persons as still foster an absurd and ridiculous belief in Ghosts, Hobgoblins, Conjurations and Enchantments.'[12] There was nothing particularly new, therefore, about invention as spectacle – or even invention for spectacle. Perkins could also look to the example of the Royal Society of Arts, founded in 1754 (though it did not receive a royal charter until almost a century later) for 'the encouragement of arts, manufactures and commerce', which held regular exhibitions of new inventions. It offered prizes for new creations, and members of the society had the privilege of 'introducing their friends on any weekday, except Wednesday, between the hours of ten and two, to examine the various models, machines, and productions, in different branches of arts, manufactures, and commerce, for which rewards had been bestowed.'[13]

Perkins would also have been familiar with the National Repository, although it might have presented an instance of the potential pitfalls, as well as the possible rewards, of technological showmanship. When the *Mechanics' Magazine* announced that a 'project is now on foot for commencing an annual public exhibition of "Specimens of New and Improved Productions of the Artisans and Manufacturers of the United Kingdom"', they were not particularly enthusiastic: 'We anticipate but little good from the project.' Their complaint – and this was clearly the patent agent Joseph Robertson talking – was that exhibition was a dangerous game for those who, like most ordinary mechanics, could not afford to patent their inventions.[14] The 'noblemen and gentlemen' behind the scheme were sure that 'it has long been a desideratum among our most intelligent merchants and manufacturers', that there should be 'an Annual Exhibition of specimens of new and improved productions of our artisans and manufacturers, conducted on a scale that should command the attention of the British Public'. It would draw attention to 'the latent talents of many skilful artisans and small manufacturers, now labouring in obscurity' – which could only be good for them, and good for the country.[15]

The National Repository held its first annual exhibition at the King's Mews in Charing Cross on 23 June 1828. It was a grand affair, with guests including 'His Royal Highness the Duke of Gloucester, Duke and Duchess of St. Albans, Marquess of Lansdowne, Marquess and Marchioness of Stafford, Earl and Countess of Jersey, Countess Ludolff, Lord Auckland, Lord Egremont', and quite a few other members of the aristocracy. The whole thing, according to the *Morning Post*, at least, was 'in the highest degree gratifying to all those who take an interest in the progress of native talent'. The exhibits offered 'striking proof of ingenuity and genius.'[16] *The Times* agreed. They were 'pleased to see that this establishment, in the Mews, Charing-cross, is becoming an object of more general attraction in proportion as the knowledge of its existence is more widely diffused'. The new exhibition was 'calculated to spread a taste for mechanical inventions among the higher classes, and thus to re-act favourably on the inventors'. It would introduce its visitors to 'a new field of knowledge and amusement of a most gratifying and lasting description'.[17] The *Literary Gazette* was full of praise for the 'various articles of curious and highly-wrought manufactures, models of looms, bridges, &c., &c., and specimens of useful and improved articles for domestic comfort or foreign commerce' that were on display.[18]

The *Mechanics' Magazine*'s initial scepticism soon turned to outright hostility, though – doubtless in part due to the prominent involvement of George Birkbeck, with whom the editors were in a state of open warfare, with the National Repository. When the exhibition reopened in 1829, the magazine was quick off the mark, criticising the quality of the exhibits and insisting that the 'protest which we published in the name of the artisans and manufacturers of the United Kingdom, against its being considered as furnishing any fair representation of their industry and skill, has received ample justification in the paltry character of the last and present exhibitions'. In fact, the National Repository was unpatriotic, and a little bit too French for comfort. Organising industrial exhibitions had, after all, been a key element in 'the plans which our arch-enemy, Bonaparte, sought to close the continent against us and bring ruin on our trade and manufactures – and that in spite of them, and of all, either enemies or rivals have done, or can do, the British manufacturer has still maintained his envied ascendancy'.[19] *The Times* soon changed its mind about industrial showmanship too. 'Is the manufacturer so degraded,' the paper enquired, 'as to require to be raised in his own estimation by seeing crowds of curious idlers, or fashionable loungers, assembled to admire his productions? Does he need the criticism of the public, who must be less skilful than himself, to improve those arts on which his existence, reputation, and fortune depend?'[20]

So, exhibition could be a dangerous strategy for an inventor, and Perkins knew it. He must have been aware of the debates raging around the National Repository and its ambitions to promote British creativity through showmanship. His own relationship with the *Mechanics' Magazine* was not entirely rosy – the editors had taken exception to some of his more extravagant claims regarding the potential of his steam gun, for example. Nevertheless, when his gallery opened, the magazine's verdict was favourable. Just as the National Repository opened its doors again in 1832 – on Leicester Square this time – they retorted that the 'importance of this unfortunate concern as an exhibition, however, is likely to be quite eclipsed by the superior attractions of the private speculation, the "National Gallery of Practical Science," in the Strand'.[21] Clearly, a great deal depended on just who was doing the exhibiting, and how. Whatever they might have thought of Perkins' credentials, and they were clearly less than impressed by the gallery's opening range of exhibits, the *Mechanics' Magazine* understood that he was his own man – he was a practical mechanic and acting independently rather than accepting patronage. Under those circumstances, exhibition appeared to be a respectable enough strategy for an inventor. But there seems little doubt that Perkins understood that his gallery could not survive simply by being a showcase for his own inventions. Dominated as the exhibits were at the beginning by Perkins' productions, he was already working to broaden the gallery's appeal by expanding the range of material on show.

Places like the National Gallery of Practical Science, and the many competitors who entered the market for spectacle in its wake, were soon to become the places to go to see what the future would be made of. There, the products of practical men and scientific gentlemen were on show, side by side, to be inspected and marvelled at. In many ways, Perkins was the inaugurator of a new tradition of scientific and technological spectacle, offering Victorian audiences tantalising glimpses of the future. It was in places like these that the Victorians learned that science could be spectacular, and where men of science and inventors learned that spectacle could be a key element in selling their inventions and discoveries. This culture of spectacle was central to the future the Victorians were in the process of inventing. Throughout the century, exhibitions kept on getting bigger, ramping up the spectacle. The Great Exhibition of 1851 put everything that had been before in the shade for sheer scale, just as it would be put into the shade by the huge international exhibitions of the century's final decades. In all sorts of ways, the Lowther Arcade was where that started.

Practical science

The National Gallery of Practical Science – or the Adelaide Gallery as it was usually called – quickly found its place on the metropolitan map of entertainment and spectacle. Visitors paid their shilling at the door to witness wonders. Even though Perkins had moved quite quickly to expand the range of exhibits in response to criticism, his steam gun remained the main attraction. At regular intervals during the day, it would be fired up to shoot a barrage of lead pellets from one end of the main gallery to the other. Visitors could watch a Jacquard loom – the inspiration for Charles Babbage's Analytical Engine – put to work. As well as electrical technology of all kinds, they might also see a hydro-oxygen microscope (a variety of magic lantern that magnified a projected image) blowing up the microscopic beasts inhabiting a drop of Thames water to monstrous dimensions. When the American natural philosopher Joseph Henry visited London to buy scientific instruments, he was delighted to see 'one of my magnets by March [*sic*] of Woolwich' on show there.[22]

The men who backed Perkins' venture into showmanship were entrenched in London's entrepreneurial and commercial circles. This was business to them. One of them was the engineer Thomas Telford. They also included Peter Harriss Abbott, the secretary of the British, Irish and Colonial Silk Company, John Strettell Brickwood of the South Sea House and wealthy financier Ralph Watson. It was these men who provided the bulk of the £16,000 needed to lease the building and purchase models and instruments. Through them, and others like them, profits made in all corners of the Empire and beyond flowed into the venture. They might have been putting up the money to help Perkins, but they were doing it to help themselves too. The Adelaide Gallery was meant to turn a tidy profit – and in the process they were putting exhibition at the heart of the business of invention. Joseph Saxton, the young American instrument maker hired to help build the exhibits, spent his time scurrying between the gallery, his own workshop in Sussex Gardens, instrument makers' shops in Clerkenwell and meetings with Brickwood at the North and South American Coffee House.[23] The gallery was a node in a network of commercial and financial interests, and a bright young man like Saxton could take advantage of that and the connections he could make. He put some of his own inventions on show at the Adelaide Gallery, like his magneto-electric engine.

What the Adelaide offered was invention as entertainment. Exhibitions of the latest thing were calculated to pull in the crowds. There were soirées and *conversaziones* for the favoured. In November 1835, for example, guests were invited 'to witness the splendid effects produced by Mr Cary's oxyhydrogen microscope, the magnifying powers of which have just been increased to the surprising extent of 3,000,000 of times'.[24] There were lectures and demonstrations laid on to explain the exhibits and their significance. 'Clever professors were there,' recalled one wag, 'who emulated the transatlantic acquirement of "knocking all creation into a cocked hat," by teaching elaborate science in lectures of twenty minutes.' It was a place where 'fearful engines revolved, and hissed, and quivered, as the fettered steam that formed their entrails grumbled sullenly in its bondage'. Visitors needed to watch out for 'artful snares laid for giving galvanic shocks to the unwary'.[25] There was a story that the Duke of Wellington had fallen into one of them and that 'the hero of a hundred fights, the conqueror of Europe, was as helpless as an infant under the control of that mighty agency'.[26] It was all a concerted and carefully calculated show of the power of invention, designed to convince its audience of the shape of things to come – and of the place of invention in that world.

Inventions that would transform the future were not the gallery's only exhibits. Visitors might also marvel, for example, at 'Weapons taken from the Native of Owyhee, who were engaged in the murder of Captain Cook'. These were not the only exotic curiosities on show. There were also 'Weapons and Tools used by the Natives of New South Wales' donated by Brickwood, as well as 'Weapons and Tools used by the Natives of New Zealand'.[27] The collection of exotics only increased over the years. Perkins' original inspiration for the gallery – Peale's Museum in Philadelphia – was similarly eclectic in its offerings. Specimens of natural history, landscape paintings and native artefacts were there along with the inventions. This was meant to celebrate westward expansion and the American ingenuity that drove it. The Adelaide Gallery was showing off invention as occupying a similar sort of role too. The knowledge on show, captured in the steam engines, electromagnetic machines and optical instruments, was knowledge that was going to transform the future. Expanding Britain's horizons and augmenting its resources was also going to be an integral part of transforming that future. The Adelaide Gallery was a showcase not just for Perkins, or for the other hopeful inventors who put their productions on display, it was a showcase for the Empire and invention's place in it.

This was invention as competition, and it was not long before the Adelaide Gallery acquired competition of its own. This came in the form of the Polytechnic Institution, which opened its doors at 5 Cavendish Square (with another door opening onto Regent Street) on 6 August 1838. This was a similar location to that occupied by the Adelaide and the intention was to attract the same kinds of crowds. The similarity was no surprise – the man behind the Polytechnic was Charles Payne, a former superintendent of the Adelaide Gallery. The money came from the property speculator William Mountford Nurse and the landowning inventor and aspiring aeronaut Sir George Cayley. There were also similarities inside, although where the Adelaide had Perkins' steam gun as its centrepiece, the Polytechnic had a full-sized diving bell. Cayley assured Charles Babbage that they had 'laid out a good round sum of money & the place by its laboratory, its theatre and its splendid Gallery is well adapted for the display of scientific discoveries & were it in truly scientific hands, so that scientific discoveries were thrown off here *hot* from the brain & before they had become public property by publication, sufficient novelty would be produced to excite public attention & to make it pay'.[28] That last remark was crucial, of course. Selling the future was meant to turn a profit.

Two shipwrights from Mumbai, who were in London to learn about English techniques of boatbuilding, thought the Polytechnic was 'quite enchanting', and that 'if we had seen nothing else in England beside the Adelaide Gallery and the Polytechnic Institution, we should have thought ourselves amply repaid for our voyage from India to England'.[29] Jehangeer Nowrojee and Hirjeebhoy

The Main Hall of the Royal Polytechnic Institution.
John Timbs, *The Year-book of Facts in Science and Art*, 1841, 3

Merwanjee had come to London under the sponsorship of John Seppings, the East India Company's surveyor of shipping in Kolkata. The two shipwrights – son and nephew, respectively, of the master shipbuilder of Mumbai, Nowrojee Jamsetjee Wadia, had come to study at the Chatham Ship Yards with the aim of bringing the latest

technologies back home with them. Jehangeer would eventually succeed his father as master builder. The family were deeply involved
in building ships both for the British Navy and the East India
Company. Seppings had 'strongly urged the necessity of taking such
a step in order that Bombay dockyards should keep pace with the
improvements of the day', and, along with rear-admiral Sir Charles
Malcolm, had recommended them to the company's board of directors.[30] The two shipwrights' fascination with the Adelaide Gallery
and the Polytechnic offers another instance of the place they, and
the spectacles they contained, occupied in the business of imperial
future-making.

These places thrived on spectacle. When rising Newcastle lawyer (and later industrialist and arms manufacturer) William George
Armstrong heard from local colliers that steam escaping from a boiler
seemed to generate large amounts of electricity, he turned the phenomenon into a spectacle for the Polytechnic. The entire exhibition
hall was shut down for a whole fortnight while Armstrong's hydroelectric machine was hoisted into place and put through its paces
before being shown off to a select audience. Consisting of a boiler
7.5 feet long and 3.5 feet in diameter, along with pipes and nozzles
to let off steam, the contraption generated impressive quantities
of electricity. The *Morning Chronicle*'s correspondent thought that
the 'passage of electricity over the tinfoil on the tubes was far more
brilliant, and the aurora borealis exceeded in intensity and beauty
anything we had ever witnessed'.[31] This really was invention for pure
spectacle's sake, designed to dazzle onlookers with the promise of
future power delivered through the concerted application of knowledge. Many years later, his fortune made, Armstrong would turn his
house and estate at Cragside, in Northumberland, into an electrical
spectacle, too. Designed by the architect Richard Norman Shaw,
best known, perhaps, for building the original New Scotland Yard

on London's Embankment, the house and ground would be largely powered by hydroelectricity.

Showing off invention was clearly lucrative enough to find a space in other exhibitions too. The Colosseum in Regent's Park featured a Department of Natural Magic that boasted the largest electrical machine in the world – a huge electrostatic generator. Built in 1827 at the eastern edge of the park, the Colosseum was another building made for exhibition. Designed with the Pantheon in Rome as its model – a deliberate invocation of empire – its main exhibit was a huge panorama of London as seen from the dome of St. Paul's Cathedral. Fashionable visitors could admire the panorama, stroll through the grounds (there were conservatories and a Swiss cottage to be enjoyed) and be astonished by scientific spectacle. Ambitious inventors hired space in exhibition halls like these to put their productions before the public. 'You did not expect to have a son turn showman, but I trust I am merely instrumental in promulgating a useful discovery,' wrote the telegraph inventor Edward Davy to his father, after putting his device on show at the Exeter Hall.[32] He recognised – and presumably hoped his father understood – that putting on a spectacle was simply part of the business of invention during this time.

In 1854, the philosophical instrument maker Edward Marmaduke Clarke – whose shop had been across the arcade from the Adelaide Gallery – opened the Royal Panopticon of Science and Art on Leicester Square. Clarke's Panopticon was an attempt to make a spectacle of invention on an even grander scale than its predecessors. It was an ambitious venture for Clarke, and a significant step up from his instrument-making business. The building, specially commissioned and designed for the purpose, was an oriental fantasy. Its 'magnificent Rotunda' was 'brilliant with colour and dazzling with glass'. There was a 'magnificent fountain' throwing a jet of water

97 feet into the air. The Rotunda's floor was 'covered with pieces of machinery, scientific apparatus, works of sculpture, a crystal cistern for diving, and many other things contrived by science for the material advantage of mankind'. One enthusiast called it 'the most magnificent temple erected for the purposes of science, of which, in modern times, we have any record'.[33] The fantasy did not last long, though. It was all a glorious failure and within a few years the building was sold off and turned into the Royal Alhambra Palace. This was a clear indication that it was possible to invest too much in showing off the future.

Outside London, the future could be a little more difficult to sell. In Manchester, a group of 'gentlemen who have been some time engaged in making arrangements for establishing a Gallery of Art and Science on popular principles', succeeded in 1839 in establishing a Practical Science Institution, soon renamed as the Royal Victoria Gallery for the Encouragement of Practical Science.[34] It was clearly modelled on the Adelaide Gallery, and its promoters appointed the London electrician and instrument maker William Sturgeon to be its superintendent. When it opened its doors to the public on 8 June 1840 in the Exchange Dining Rooms, its 'models and specimens of art and science' were 'so stamped with ingenuity, industry, and talent, as to deserve the utmost attention from all classes of our townspeople'. *The Manchester Times* boasted that 'the exhibitions of the Polytechnique and Adelaide Gallery are greatly inferior to that of our Victoria Gallery in everything practical and useful'.[35] A couple of years later, the gallery moved to rooms in Manchester's Royal Institution, where it continued 'to provide a *permanent* collection of models and apparatus to illustrate the arts and manufactures; and for the display of experiments combining philosophical instruction with general entertainment'.[36] The move was not a success. The gallery closed soon after, leaving Sturgeon effectively penniless.

The Polytechnic, however, continued to flourish. Throughout the 1850s and 1860s, under the management of John Henry Pepper, Londoners flocked there to see the future of invention. He made outsize, spectacular exhibition his trademark. In 1869, Pepper introduced the monster coil – a massive induction coil designed and built by the instrument maker Alfred Apps, who had started making a name for himself as a maker of electrical apparatus a few years earlier. Newspapers dwelled on the coil's dimensions. It was almost ten feet long and two feet in diameter. It weighed in at fifteen hundredweight (three-quarters of a ton). The wire in its secondary coil was 150 miles long. It produced 'a spark, or rather a flash of lightning, 29 inches in length and apparently three-fourths of an inch in width, striking the disk terminal with a stunning shock'.[37] The gigantic instrument 'allows all the destructive phenomena of chamber electricity to be exhibited, in hitherto unapproached beauty and intensity'. The coil itself 'displays the name and address of its Mr Apps, its maker, in gold letters of considerable size'. Sometimes, 'the attraction of these gold-leaf surfaces was sufficient to divert the spark from its course, and visibly to break it up into portions'.[38] This branding might have reminded the audience that though it might be Pepper performing on stage, the coil itself was the product of considerable skill, labour and resources.

In many ways, that was what these exhibitions of spectacular invention were about. Certainly, they celebrated spectacle. They glorified the individual inventor turned showman as the generator of innovation and progress. They promoted the idea that invention was the product of individual ingenuity. It was Perkins' steam gun, Armstrong's hydroelectric machine or Davy's telegraph that spectators came to wonder at, after all. But at the same time, the devices those spectators saw were the result of the hard-won and expertly applied skills of small armies of people. Behind the scenes at the exhibition, a great deal of work was going on. As inventor–entrepreneurs competed with each other to produce ever more elaborate and extravagant exhibitions of their inventions, the amount of work invested in them, and the resources they required for their mounting, expanded too – sometimes to unsustainable levels, as Edward Marmaduke Clarke's over-ambitious Panopticon suggests. At places like the Adelaide Gallery and the Royal Polytechnic, the future was being made visible – and it was a future that depended on the systematic orchestration of people and resources on an entirely unprecedented scale. Pepper's monster coil was made not just by Alfred Apps in his instrument maker's shop on the Strand in London, but with copper, cotton, glass, gutta-percha, iron and wood transported from all corners of the Empire and beyond.

Great exhibition

Nowhere was the imperial reach of scientific and technological spectacle more apparent than at the Great Exhibition in Hyde Park in the summer of 1851. The huge pavilion of iron and glass was itself a statement about power and the future, and its contents were the stuff of tomorrow, as well. It had been designed by the unlikely figure of Joseph Paxton. Paxton had established his reputation as the Duke of Devonshire's head gardener at Chatsworth in Derbyshire, where, from the 1830s onwards, he had designed and built a series of greenhouses of increasing size and complexity. The Great Conservatory he built for the duke in 1836 was the largest glass building in the world at the time and was constructed using special glass plates made by the glass manufacturer Robert Chance. The design he offered the commissioners for the Great Exhibition was in effect a gargantuan greenhouse, and its building depended on the latest innovations in cast iron and glass manufacture. 'The Exhibition is great in its design; great in regard to the beauty and size of the building in which it is to take place; great as respects the number of persons it will bring together; and greater than all in its probable effects on society at large,' the *Child's Companion* told its readers.[39] It was not long before it was dubbed the Crystal Palace. It was 'an Arabian Night's structure, with a certain airy unsubstantial character about it which belongs more to an enchanted land than to this gross material world of ours'. It looked like a 'splendid phantasm' that might dissolve at any moment, or fade back into the London mist.[40]

The exhibition had its origins a few years earlier. Looking at industrial exhibitions on the other side of the English Channel, Henry Cole decided that he could do better. Cole, who had worked with Rowland Hill in establishing the Penny Post and introduced the first commercial Christmas cards a few years later, was a leading member of the Society for the Encouragement of Arts, Manufactures and Commerce. He was also a friend of Prince Albert and used his

Interior of the Great Exhibition of 1851.
Great Exhibition of the Industry of All Nations, 1851: Illustrated Catalogue
(London: Bradbury and Evans, 1851)

influence to secure a royal charter for the society in 1847. Cole used his position as a council member of the society – and his association with Prince Albert – to lobby for the idea for the Great Exhibition, building on small but successful exhibitions organised by the society in 1848 and 1849. The result of all this assiduous politicking was the establishment of a royal commission in 1850, with Albert at its head, to organise an exhibition on a grand scale the following year. Invitations were sent out to other countries inviting their participation. This was going to be the Great Exhibition of the Works of Industry of All Nations – though few (if any) of the organisers really doubted that it would be an exhibition of British industrial supremacy. Plans were laid for organising and classifying the exhibits. Charles Babbage argued (unsuccessfully) that the exhibition would be meaningless unless each exhibit carried a price tag – displaying the price, he insisted, was essential for promoting industrial competition. A competition was also established to design a building, which would need to be both impressive and temporary.

Paxton's design won because it fulfilled those essential requirements – and because Paxton organised a successful campaign in the press in support of his proposal. The *Illustrated London News* championed his proposal, in particular, publishing the plans even before they had been approved by the commission. It was vital that the Crystal Palace be erected quickly and be taken down just as efficiently at the end of the exhibition – and with as little effect on Hyde Park itself as possible. Paxton even succeeded in using his gardener's experience to allow for some of the park's trees to stand inside the building. The building was designed in prefabricated sections to be assembled on site by skilled workmen. It was difficult – and sometimes dangerous – work, scrambling over the huge pavilion's iron skeleton fitting glass sheets into grooves. At the end of November 1850, the glaziers went on strike. Fully aware of their value, they demanded higher wages than the four shillings they were paid, as well as a reduction in the pace of work – they were expected to fit 58 panes of glass a day to earn their money. Despite the fact that the government hoped that the exhibition would be a symbol of renewed unity after the nation's flirtation with revolution during the Chartist uprisings in 1848, the authorities came down hard on those involved. The strike was broken and the leader, William St Clair, arrested and charged.

The Great Exhibition of the Works of Industry of All Nations opened with due pomp and circumstance on 1 May 1851. Queen Victoria herself led the proceedings in weather 'as genial and bright as if there had been no holiday to be spoilt by its being otherwise'. 'It is no exaggeration to say,' said the *Daily News*, 'that the spectacle within the Crystal Palace far exceeded, in impressive beauty, what the most sanguine could have anticipated.' The Queen 'entered into her part with a heartiness and vivacity that carried the willing sympathies of all around her'. The Duke of Wellington 'contemplated the scene with as much of pleased vivacious interest as if it had been his twentieth instead of his eighty-second birthday'. The glittering throng of courtiers, diplomats and politicians was testament to just how freighted with significance the exhibition already was. 'There never was before an industrial exposition to which so many remote and varied lands and communities had sent their contributions,' declared the *Daily News* – and there 'never before was one in which so much of mere material wealth, enhanced by the immense amount of inventive, persevering intelligence and industry to which it bore testimony, was accumulated'.[41] This really was the future – and the forces that were in the process of shaping the future – on show.

The organisers had been assiduous in attracting exhibits and exhibitors from Europe, the Empire and beyond, and the place was crammed with things to see and admire. Inside the Crystal Palace, the space was organised like a cathedral of industry. Entering from the west, visitors found themselves looking down a long nave towards a central fountain where the nave and transept intersected. Like side chapels in a cathedral, there were galleries all along the nave and further galleries on an upper floor. As they entered, most of what visitors saw around them were products of Britain and the Empire's factories and workshops. Most European countries' industries were to be found on the further side of the central transept. The overwhelming impression was of crowds and colour. The Crystal Palace was packed daily throughout the summer of 1851. Even Derby day – a key date in the fashionable London season – 'made little impression on the appearance of the crowd at the Crystal Palace'.[42] The organisers had to institute a one-way system to keep the crowds moving. On shilling days (the usual price for admission was five shillings) commentators drew favourable comparisons between the earnest attention devoted by artisans and mechanics to the exhibits themselves, as compared to fashionable loungers who were only there to be seen.[43]

Whether they were fashionable loungers or earnest artisans, the visitors who poured into the Crystal Palace found themselves surrounded by novelty. Scale was one thing – commentators dwelled on the sheer number of things the exhibition contained. There were examples of textiles manufactured by the latest machinery in Britain – and by hand in India. There were specimens of raw material from around the world side by side with the commodities manufactured from them. There were steam engines, steam hammers and steam presses at work. Visitors were impressed by telegraphs and electrical apparatus of all sorts. Particularly impressive was the electric clock, designed by Charles Shepherd, that was built into the fabric of the building itself above the western entrance. All other clocks in the building were synchronised with it through telegraph wires. A remarkably clear daguerreotype of the Moon taken by the Harvard astronomer William Cranch Bond and John Adams Whipple caused a sensation. The curious could examine examples of the latest domestic appliances and ornaments – there was even a stove made to look like a coat of medieval armour. As they worked their way through the throng, visitors were assailed by sights and sounds. Among the artefacts of empire on show was the recently acquired Koh-i-Noor diamond, looted after the annexation of the Punjab in 1849, thought to be the largest in the world, displayed to the exhibition's visitors imprisoned in a gilded cage.[44]

Many visitors to the Crystal Palace came and came again – the Queen was a frequent visitor, as, unsurprisingly, was Albert. Charles Babbage was another one, despite his disapproval of some of the arrangements. It became a habit for many of the metropolis' gentlemen of science to take visitors to the exhibition to enjoy the improving show. Charlotte Brontë went twice during a visit to London. She thought it was 'a wonderful place – vast, strange, new and impossible to describe. Its grandeur does not consist in one thing, but in the unique assemblage of all things. Whatever human industry has created you find there, from the great compartments filled with railway engines and boilers, with mill machinery in full work, with splendid carriages of all kinds, with harness of every description, to the glass-covered and velvet-spread stands loaded with the most gorgeous work of the goldsmith and silversmith, and the carefully guarded caskets full of real diamonds and pearls worth hundreds of thousands of pounds.' 'It seems,' she described, 'as if only magic could have gathered this mass of wealth from all the ends of the earth – as if none but supernatural hands could have arranged it thus, with such a blaze and contrast of colours and marvellous power of effect.'[45] Charles Dickens, on the other hand, thought that it was all a bit too much, though his journal *Household Words* paid the exhibition plenty of attention.

The Crystal Palace closed its doors for the last time on 15 October. It had been a success beyond the wildest imaginations of its organisers. During that summer, some 6 million people visited the Great Exhibition. On one single day in the final month, more than 100,000 people passed through it. It was testimony to the degree to which imaginings of a future underpinned by technological innovation had captured the Victorian middle classes. Shortly afterwards, work began on dismantling the Crystal Palace itself. The building was not destined for the scrapheap, though. Its components were carried across London to Sydenham, where it was all reassembled into a building even more spectacular than before. For the rest of the century and beyond, it remained a showcase for invention, an embodiment of technological spectacle, until it burned to the ground in 1936. The 'beautiful building glistening like a fairy pavilion in the sunshine, and the spreading grounds to ramble in', became part of London's cultural landscape. Visitors rambling through the grounds would stumble over a menagerie of dinosaurs, designed by sculptor Benjamin Waterhouse Hawkins. Made to the latest scientific specifications of how such beasts would have appeared, the dinosaurs were there to remind visitors of the degree to which the Victorians' scientific and technological grasp dominated the deep past as much as it did the coming future.

William Whewell, gentleman of science, polymath and master of Trinity College, Cambridge, offered a particularly powerful – and telling – view of what the exhibition had done. It had turned the march of progress into a frozen tableau that transcended time. All nations progressed, but some had progressed more than others. What the Great Exhibition had done was to bring them all together so that they could be compared side by side. 'Different nations have reached different stages of this progress, and all their different stages are seen at once, in the aspect which they have at this moment in the magical glass, which the enchanters of our time have made to rise out of the ground like an exhalation,' Whewell suggested. By bringing the exhibits all together and showing them in the same spectacular place, and 'annihilating the space which separates different nations', the exhibition's organisers had produced 'a spectacle in which is also annihilated the time which separates one stage of a nation's progress from another'.[46] The Reverend George Clayton sermonised that the 'repository of wonders', that made up the Great Exhibition, 'may be regarded as a BENEFICIAL STIMULUS TO HUMAN DILIGENCE AND INDUSTRY. To those who have paid due attention to the constitution and qualities of our race, it must often have occurred as an unspeakable advantage, that occupation is everywhere supplied to the great masses of mankind, both as to their corporeal activities and their mental energies'.[47]

Very much on display at the Great Exhibition – and this in many ways was the chief burden of Whewell's commentary – was imperial reach and power. Walking through the Crystal Palace, visitors saw what empire offered. It was an unprecedented feat of organisation and the marshalling of resources that could only have been achieved through the exertion of imperial power. The presence of the Koh-i-Noor as a spoil of empire simply served to crystallise what was already clear. The contents of the Crystal Palace demonstrated that the technological innovation that Victorians dreamed would generate new tomorrows for them was entirely dependent on the power to acquire and deploy resources – human and material – on a global scale. Whewell did not (quite) say it, but everyone understood and celebrated the fact that the Great Exhibition was an exhibition of British dominance. Cartoons in *Punch* depicted the Crystal Palace sitting literally on top of the world. The satirical magazine was poking fun at the organisers' pretensions, of course, but the joke captured the reality. The Great Exhibition was a roaring success – and the exhibitions that followed it throughout the following half-century were roaring successes (though not always repeating the Great Exhibition's financial bonanza) – because the Victorians really were on top of the world by the middle of the nineteenth century. They were rushing headlong towards the future that the spectacle of science offered them.

Technological sublime

The Crystal Palace's success offered an object lesson on the place that science and invention had come to occupy in Victorian national culture by the middle of the nineteenth century. Looking back from the vantage point of the 1890s, the *Electrical Review*'s editor would speculate that it 'would be interesting if we could know how the future historian will deal with an institution which is peculiar to the nineteenth century'. This was the age of exhibitions, they thought. From the middle of the century onward, 'we have had International, General and Special Exhibitions of all kinds. Bazaars and marts are old enough, but an exhibition, though allied to both, is neither one nor the other, and no preceding institution will be found to exactly compare with it. The historian will probably come to the conclusion that the institution existed in the latter half of the nineteenth century, because it was one suited to the requirements of the period'.[48] The simple fact that exhibitions like these were so popular, and so universal, underlined the increasing conviction that innovation was the engine of progress. Exhibitions were also themselves powerful symbols of national power and prestige. Organising an international exhibition sent a powerful message to the world that this was a state with the capacity and the resources to embrace the future.

About £300,000 of the profits from the Great Exhibition paid for the building of the Great London Exposition held in the summer of 1862. Held in South Kensington, like its predecessor, the exposition was meant to dwarf the Great Exhibition both in scale and in scope. 'Uplift a thousand voices full and sweet, in this wide hall with earth's invention stored,' sang the choir at the opening ceremony. These words by poet laureate Alfred Lord Tennyson simply emphasised the degree to which the appreciation of innovation was now ingrained in Victorian culture. It proved just as popular as its predecessor too. 'There never was such an invasion of the west before,' remarked one newspaper on opening day.[49] By the time the exhibition closed its doors for the last time on 1 November, more than 6 million visitors had passed through them. Those visitors would have been counted in and out of the exhibition building by the electric turnstile patented by telegraph inventor Charles Wheatstone. They would have been able to marvel at a large chunk of Charles Babbage's unfinished Calculating Engine. The Lady of the Lake, one of the London and North Western Railway Company's latest design of steam locomotive, was another star attraction. Implicit in Tennyson's ode to the exhibition was the notion that the whole show underlined the relationship between Britain's imperial reach and its pursuit of a technological future.

Britain, of course, was not the only European empire in pursuit of such a future. Paris responded to the challenge with its own *Exposition Universelle des Produits de l'Agriculture, de l'Industrie et des Beaux Arts* (Universal Exhibition of the Products of Agriculture, Industry and the Fine Arts) in 1855. It was a deliberate attempt by the recently crowned Emperor Napoleon to assert France's hopes for an imperial future. The next Parisian exhibition, the *Exposition Universelle d'Art et d'Industrie* in 1867, was even bigger and more ambitious. It was held on the Champ de Mars and its grounds covered more than

170 acres. The Suez Canal Company had its own pavilion where visitors could marvel at models of this great feat of French engineering and listen to Ferdinand de Lesseps give lectures on what it meant for the future of the Second French Empire. Crowds lined up to try out the hydraulic elevator. The latest in electromagnetic engines and generators was on show and provided inspiration for a visiting Jules Verne (they would feature in *Twenty Thousand Leagues under the Sea*, for example). Another star attraction was an enormous cannon built at the Krupp ironworks in Prussia. It weighed 50 tons and could fire 1,000-lb cannonballs. The Krupp cannon was back in Paris a few years later, pounding the city's defences during the siege as part of the Franco-Prussian War.

In 1873, the Vienna International Exhibition signalled the Austrian Empire's ambition to join the race towards the future. It was hailed as 'a work which draws upon Austria the eyes of the world, and secures to our Fatherland the recognition of prominent participation in the promotion of the welfare of mankind by instruction and labour'.[50] It was 'the great event that was to proclaim New Austria the peer in hopeful enterprise and self-improvement, of her elder sisters, England and France'.[51] Again, it was a huge affair, the exhibition grounds spread out over almost 600 acres. The main building was more than half a mile in length. There were steam-driven electromagnetic generators in the Machine Hall, including the latest dynamo designed by the Belgian inventor Zénobe Gramme. A few years later, the Centennial Exhibition held in Philadelphia in 1876 signalled the United States' ambitions for the future. New York had held its own Exhibition of the Industry of All Nations in 1853, with its own New York Crystal Palace, but that had been a relatively small affair. The Centennial Exhibition was on a far larger and more ambitious scale. The Main Exhibition Building, enclosing over 21 acres, was the largest building in the world. The Machine Hall housed the gigantic 1,400 horsepower Corliss Centennial Steam Engine that powered every machine in the hall. Visitors could travel around the exhibition grounds on the steam-driven Centennial Monorail. Another exhibit was Alexander Graham Bell's telephone.

In 1880, the colonies joined in, with the Melbourne International Exhibition in the Australian province of Victoria. Planning had been under way since 1878, and rival city Sydney attempted to steal Melbourne's thunder with an exhibition in 1879, but it ended up being relatively small by comparison. Melbourne's Main Building covered seven acres and was crowned by a dome 60 feet in diameter and rising to more than 200 feet above the building. The exhibition was ceremonially opened on 1 October by the Marquess of Normanby, the governor of Victoria, who declared that any 'country might be proud of such a display, and of the foresight, energy, and ability manifested by all concerned in the work'.[52] A few years later, in 1887, the colony of South Australia held the Adelaide Jubilee International Exhibition to celebrate the 50th anniversary of Victoria's coronation, with a railway line running directly from the city station to the exhibition itself. Two years later, Melbourne's Main Building was refurbished and made even bigger to host the Melbourne Centennial Exhibition celebrating 100 years of British colonisation in Australia. Exhibitionism was a way of demonstrating that even Britain's most distant colonies fully participated in the imperial drive towards the future, and that they possessed the qualities needed to get there.

In 1881, Paris held its *Exposition Internationale d'Électricité* at the Palais de l'Industrie on the Champ de Mars – built originally for the *Exposition Universelle* in 1855. It was a huge affair: 'Engines amounting in the aggregate to over two thousand horse-power will supply such a number of electric lamps as never have been accumulated within any building nor in any place, and, grand as are the dimensions of the Palais de l'Industrie, the brilliancy of the illumination must necessarily be amazing.'[53] Visitors flocked to see the latest in dynamos and electric lights, and the sheer scale of the exhibition simply served to underline the fact that electric illumination was by now about more than spectacle – it was becoming a serious commercial prospect. The electrical engineer Marcel Deprez showed off his system of long-distance DC electrical distribution. Outshining it, though, was Thomas Alva Edison's exhibition of his new system. It was 'certainly the most noteworthy object in the exhibition', according to William Henry Preece, the Post Office's chief telegraph engineer.[54] This was the exhibition that played host to the first meeting of the International Congress of Electricians, where a battle royal was fought out over the thorny issue of electrical standards and units.

The Crystal Palace hosted its own International Electrical Exhibition the following year. *Punch* poked gentle fun at it all, caricaturing a parade of electrical technology sweeping chimney sweeps, gas lighters and coal merchants away into a dusty past while electricity marched into the future. It was satire that acknowledged the coming force, though. As the exhibition opened, newspapers predicted that ''ere long the palace will in the evenings be a competitive competition of electric lighting, being at the same time the most complete exhibition of electric lighting that has as yet been seen'.[55] The Crystal Palace hosted another Electrical Exhibition a decade later in 1892. It would offer a spectacle of a coming electrical age where the 'telegraph and the telephone, the phonograph and the electric light, have become almost necessaries of life, and nearly every day sees some new application of the mysterious fluid to the ordinary work of the world'.[56] Following his own triumphant performances at the Royal Institution at the end of January, Nikola Tesla, that coming man of the future, featured in the exhibition too. Soon afterwards, the

Crystal Palace's proprietors were advertising that Tesla's spectacular apparatus was now part of the show.

The *Exposition Universelle* held in Paris at the Champ de Mars over the summer of 1889 had as its centrepiece the Eiffel Tower – the tallest building in the world. Built of iron girders by the civil engineer Gustave Eiffel, and largely illuminated by electricity, the tower was meant to symbolise France's drive towards the future. It was 'a credit to the energy and constructive skill of Frenchmen', said the English press.[57] 'The International Exhibition, which is to open on 6 May at Paris, exceeds in dimensions anything that has yet been seen of Exhibitions in that city,' they reported. It would 'exceed anything that has gone before it'. A great deal was made of this sheer scale – a recognition that the vast quantity of resources required to assemble the whole thing was an essential aspect of the show.[58] The Eiffel Tower was not the only outsized edifice that featured. The massive Galérie des Machines covered twenty acres of ground. Built of iron, glass and steel, its vaulted arches spanned 377 feet. The opening ceremony was pure spectacle. 'The whole of the Exhibition buildings are brilliantly lighted up. The two rows of fountains leading up to the vast *jet d'eau* in the centre of the gardens produced wonderful effects by the changes of illuminated colour they were made to undergo by an instantaneous process.' The Eiffel Tower outshone everything else though: 'Vast flashes of electric light were distributed over Paris from a revolving lantern tower, giving to the green avenue of the Champs Élysées the appearance of pure silver when chancing to fall on them.'[59]

Size and spectacle were also the defining features of the Columbian Exposition in Chicago in 1893. Just as Paris in 1889 celebrated the storming of the Bastille, the Columbian Exposition celebrated four centuries of North American colonialism. The exhibition site, designed by the fashionable Boston landscape architect Frederick

Detail of the Eiffel Tower at the Paris Exhibition in 1889.
G. Hentschel, 'La Tour Eiffel' (1889), *Paris: Capital of the 19th Century,*
Brown Digital Repository, Brown University Library

Law Olmsted, spread out over 630 acres on the shores of Lake Michigan, laid out around a network of canals and lagoons. At its epicentre, the Manufactures Building covered 30 acres. The site even had its own railway station and an electric walkway along the lake's shore. Chicago's answer to the Eiffel Tower was the Ferris Wheel, 264 feet high and powered by enormous steam engines. Inside the Electricity Building there were spectacular displays by rival Edison and Westinghouse companies, including the Tower of Light – 'a glorification of the Edison lamp and the Edison system of incandescent lighting'.[60] It was 'a magnificent triumph of the age of Electricity'.[61] It was a 'shining vision, serenely awaiting the admiration of the world'.[62] Patriotic Americans were exhorted to 'make a careful general survey of the Fair itself, as illustrating the present condition of our nation from many points of view, and likewise its promises and prospects for the future'.[63]

A view of the Columbian Exposition.
C. D. Arnold and H. D. Higinbotham, *Official Views of the World's Columbian Exposition* (Chicago: Chicago Photo-gravure Co., 1893)

From the 1830s onwards, and increasingly so throughout the second half of the century, exhibition culture became vital to the business of invention. It mattered not just because exhibition helped to sell inventions to a public greedy for the future. It mattered because exhibitions came to define what invention was and what inventions were for. They offered the visitors that flooded through their gates a tangible glimpse of a future made out of the products of invention. The popularity of exhibitions of invention throughout the Victorian age – and their increasing prominence and scale towards the century's final decades – speaks to the accuracy of the *Electrical Review*'s remark that this really was the 'age of exhibitions', and that these huge celebrations of invention were indeed 'suited to the requirements of the period'. For inventors, they offered a road to fortune – the *Electrical Review*'s recognition of their descent from bazaars and marts acknowledged that their role was to provide a marketplace for invention. By doing that they turned inventors into heroes too. As far as the crowds that flocked to see their engines and systems on display were concerned, men like Edison, Gramme or Siemens were the real makers of the future. But nations jostled to claim ownership of the future too.

More than anything else, exhibitions showed off the resources that were needed to generate the future. By the end of the century, organising an exhibition needed time, commitment and the capacity to deploy large numbers of people quickly and efficiently – things that only states and large corporations could manage. They were important symbols of national power. As such they were vital elements in forging the modern state. Running through exhibitions on every level was the interrelationship between expertise and future-making. They might make heroes out of individual inventors, but they also laid bare the fact that exhibitions – and the business of invention – were mass productions. Putting on a good show required resources on a global scale. Exhibitions were demonstrations of the imperial reach of the nations that hosted them. To ram that message home to their audiences, by the end of the century, international exhibitions typically featured displays of the native peoples of their hosts' colonies, juxtaposed with their displays of the latest technological innovations. They were graphic reminders of William Whewell's suggestion that exhibitions could be understood as presenting frozen tableaux of the rise of civilisation. In the end, the view of the future offered at these exhibitions was one thoroughly intertwined with imperial ambitions.

Chapter 5

Fuelling the Future

I t must have been an astonishing spectacle. For six days in October 1829, five behemoths competed with each other for the future of the railways. The competitors – four different steam locomotives as well as one driven by horsepower – were fighting for the right to pull the trains on the soon to be opened Liverpool and Manchester Railway – and a prize of £500 (about £60,000 in modern money). The steam locomotives taking part were the *Novelty*, the *Perseverance*, the *Rocket* and the *Sans Pareil*. The final competitor, the *Cycloped*, was driven by a horse walking along a perpetual driving belt. This was serious business and all the competitors recognised just what was at stake. The trials would decide whether steam really was going to be the future of locomotion. Triumph would seal the victor's fortune as well – the 'offer of so handsome a premium, and the brilliant professional prospects which the winning of it presented to mechanical men, excited a very spirited competition among them'.[1] It was no surprise that the machines entered for the competition were the latest technological wonders. Their designers were all experienced men with impressive track records as innovating engineers, and with an intimate knowledge of the workings of rail. Thundering back and forth along a mile of track at Rainhill over the six days of the competition, the winner would show the world what the genius of practical men might achieve. This was engineering at the edge.

The Liverpool and Manchester Railway was itself an ambitious and forward-looking project. It was the first railway built to connect two cities and intended to carry passengers as well as goods. Carrying raw materials from the thriving port of Liverpool to the Lancashire cotton mills around Manchester, and finished products back again, it would be a direct challenge to the monopoly of the canal system. The railway would be 'a means of conveyance manifestly superior to existing modes: possessing, moreover, this recommendation in addition to what could have been claimed in favour of Canals, namely, that the Rail-road scheme hold out to the public not only a cheaper, but far more expeditious conveyance than any yet established'.[2] Unsurprisingly, the canal owners (including the Duke of Bridgewater, whose land the railway would need to cross) hated the idea, and succeeded in blocking the proposal for a while, but work on the line eventually started in June 1826. It was central to the Liverpool and Manchester Railway Company's plans from the outset that their railway would run on steam. No horse-drawn carriage would be allowed on their tracks. The original plan was for locomotives to be pulled back and forth by stationary steam engines, but George Stephenson, the company engineer, argued for steam locomotives instead. The Rainhill competition would decide the issue.

Stephenson's plan was a bold one – typical of the practical man's approach, and of his own dreams of the future. Competition would show the world the superior machine, and the superior man. It would make sure that the world knew about the ambitious plans for the Liverpool and Manchester Railway, too. It was going to be the beginning of 'a new epoch in the progress of mechanical science, as relating to locomotion.'[3] The trials were designed to be a test of speed and stamina. The engines would be tried to the very limits of their capabilities. The outlier the *Cycloped* was the first to fail when the horse powering it fell through the carriage floor. The *Perseverance* failed to attain the minimum speed of ten miles an hour. The *Sans Pareil* cracked a cylinder. The *Novelty* was the favourite and succeeded in achieving 28 miles an hour before one of its pipes burst. The only one left standing by the end of the trials was the *Rocket*. It had been 'altogether a new spectacle, to behold a carriage crowded with company, attached to a self-moving machine, and whirled along at the speed of thirty miles an hour.'[4] The triumphant engine had been designed by Robert Stephenson, George Stephenson's own son. It was a triumph for the Stephenson dynasty, as well as a triumph for steam.

It seemed clear to everyone that steam now really would fuel the future. Steam would power the pursuit of progress. The confident predictions of its promoters – men like the Stephensons – were ripe for satire, though. Not everyone was happy to participate in the mad dash into the future that steam seemed to promise. The transformative power of technology did not convince everybody. Between 1825 and 1829, for example, the caricaturist William Heath drew a series of satirical prints – all titled *March of Intellect* – that poked fun at the growing cult of progress. Steam made for an easy target. One cartoon featured 'The Steam Horse Velocity'; a steam coach that could travel between London and Bath in six hours; and the Grand Vacuum Tube Company's contrivance to travel 'Direct to Bengal' from the summit of Greenwich Hill. There were balloons and airships in the sky everywhere, of course. The world Heath depicted was one in which steam-powered technology had run amok. But reality was in the process of overtaking these satirical fantasies. The *Rocket's*

Rainhill victory ensured that steam locomotion really would be the chosen technology of the Liverpool and Manchester Railway. The line's commercial success over the next few years made their choice the obvious one for others to follow as the railways expanded in the coming decades.

Thirty miles an hour may have been the stuff of dreams in 1829, but within a little over a decade, steam locomotives were thundering along at 50 miles an hour or more. The railway network kept on growing, and by the 1840s fortunes were being made and lost in railway mania. Steam locomotion and the railways seemed made for each other. The 'wheels, rails, and carriages are only parts of one great machine, on the proper adjustment of which, one to the other, entirely depends the perfect action of the whole'.[5] Dionysius Lardner, the outspoken and radical professor of natural philosophy at the London University, prolific commentator on all things scientific and editor of the *Cabinet Cyclopaedia*, thought that the 'moral and political consequences of so great a change in the powers of transition of persons and intelligence from place to place are not easily calculated'. The spread of steam meant that the 'concentration of mind and exertion, which a great metropolis always exhibits, will be extended in a considerable degree to the whole realm'.[6] The steam-driven railways seemed to perfectly capture the Victorian drive into the future – and they seemed ideal expressions of a gung-ho engineering culture. They made their passengers see the world differently. They even gave them a new kind of disease (railway spine, supposedly caused by being rattled around so violently as the locomotives sped along the tracks) as they were hurtled on towards their destinations.[7]

All this makes it even more peculiar that by about 1850 many Victorians seemed to think that steam was a technology already from the past, rather than for the future. The future – and the future of locomotion in particular – was going to be electric. 'I cannot discover any good reason why the power may not be obtained and employed in sufficient abundance for any machinery – why it should not super-cede steam, to which it is infinitely preferable on the score of expence, and safety, and simplicity,' said the *Morning Herald*. 'Half a barrel of blue vitriol, and a hogshead or two of water, would send a ship from New York to Liverpool.'[8] Or as the Welshman William Robert Grove put it to the London Institution at the beginning of the 1840s: 'Had it been prophesised at the close of the last century that, by the aid of an invisible, intangible, imponderable, agent, man would in the space of forty years, be able to resolve into their elements the most refractory compounds, to fuse the most intractable metals, to propel the vessel or the carriage ... the prophet, Cassandra-like, would have been laughed to scorn.'[9] The end of the age of steam and the beginning of an electrically driven future was just a matter of time, it seemed.

Steam led the way into the future for the Victorians, but, just as clearly, that future belonged to electricity. The possibilities – and the limitations – of a steam-driven culture raised new questions about the shape of the future and how it would be fuelled. As far as some commentators were concerned, the capacity to harness the powers of nature and put them to work was an important index of civilisa-tion. By this kind of argument, the ability to control steam – and the future ability to control electricity – was what made Victorian culture exceptional. Victorians were well aware that the resource that fuelled their contemporary steam-driven culture was a finite one. The coal from the South Wales valleys would eventually all be consumed – one reason, at least, why the future would not be in steam. Victorians were on the alert for new ways of powering society – for illuminating their streets and houses, for running the machines that churned out cheap goods in their factories and the engines that rushed them from place to place. That was what the exhibitions they flocked to throughout the second half of the century showed them, and what the culture of scientific precision and accuracy seemed to promise too.

Magnificent apparatus

It began with William Sturgeon's invention of the electromagnet in 1825. Born in Lancashire in 1783, Sturgeon had been unhappily apprenticed to a shoemaker at the age of thirteen, before joining the Westmorland Militia in 1802 and the second battalion of the Royal Artillery a few years later. Unhappy his life as an apprentice may have been, but it had provided him with some mechanical skills and a knowledge of machinery that he developed for himself during his time as a gunner. Posted to Newfoundland, he traded his mechanical knowledge and boot-making skills to officers in exchange for their turning a blind eye to his efforts at self-improvement. He taught himself mathematics, as well as Greek and Latin. According to his own account, it was in the aftermath of a particularly violent thunderstorm that he developed the fascination with electricity that would dominate his future life as an experimenter. Back at Woolwich, he used the skills he had acquired both as an apprentice and in the army to start making a name for himself as a maker of electrical instruments. Thanks to the patronage of some of the Royal Military Academy's professors, such as Peter Barlow, he found work lecturing at the East India Company Military Academy at Addiscombe. Humphry Davy at the Royal Institution was less impressed by someone of Sturgeon's class presuming to be a man of science. He recalled bitterly to the American man of science Joseph Henry how he 'first became dissatisfied with the Royal Institution by a harshe remarke of Sir H. Davey who when Mr S. shewed him some exp. On magnetism said he had better mind his last than be dabling in science'.[10]

Like many practical electricians, Sturgeon saw the world in straightforward terms of the ebb and flow of the electrical fluid.

The Universe was made up of bits and pieces of electrical apparatus just like the ones he made in his workshop. Indeed, the electrician's task was to 'mimic nature's operations', and try against the odds to reproduce something like 'the magnificent apparatus of the earth'.[11] In 1824, for example, he built a sphere from silver and platinum wires around a magnet, which rotated when electricity was generated as the metals were heated – the idea was that this provided a mode of how the Earth itself might rotate on its axis.[12] For much of his life scientific after returning to England in 1820, Sturgeon scraped out a living as an instrument maker and lecturer. For someone like him, his intimate and practical understanding of the way his electrical instruments worked was central to his sense of self – and proof of his knowledge of nature's workings too. This was a robust and no-nonsense view of things that emphasised the hard-earned skills of the practical man. In 1825, Sturgeon presented the Society of Arts with a set of electrical instruments designed 'to illustrate the subject in public lectures'.[13] One of those instruments was the electromagnet, made to magnify magnetic power, but with the important additional feature that it allowed the magnetic force to be rapidly switched on and off. What that meant – as Sturgeon and other practically minded electricians soon realised – was that electromagnets could be used to make electromagnetic engines.

Michael Faraday's discovery of electromagnetic induction in 1831 caught the attention of these practical electricians too. While tinkering with wire coiled around a cast-iron ring, Faraday noticed that whenever a source of electricity was connected or disconnected to one coil around the ring, a current of electricity would also flow through the other coil. Similarly, when a magnet was moved in or out of a coil of wire, it would produce electricity. Within a few months of Faraday's discovery, the American instrument maker Joseph Saxton, working for Jacob Perkins at the Adelaide Gallery, had invented

William Sturgeon's electromagnet takes central place
in the frontispiece of Henry M. Noad's *Lectures on Electricity*
(London: George Knight, 1844).

what he called a 'magnetoelectrical machine' based on this principle. The machine could produce a continuous current of electricity. At about the same time, the French instrument maker Hippolyte Pixii developed a similar device. On 14 November 1833, the Adelaide Gallery hosted a contest between the two devices to see which one produced the most spectacular performance. While Pixii's machine 'rapidly decomposed water' with its current, Saxton's machine 'gave powerful shocks, brilliant sparks, heated a platinum wire red hot, and decomposed water'.[14] Saxton won this first bout, but he soon had more competition. The instrument maker Edward Marmaduke Clarke, whose shop was a stone's throw away from the Adelaide Gallery, had invented his own variation. This was not a machine for the timid. Clarke boasted that the 'effect this armature produces on the nervous and muscular system is such that no person out of the hundreds who have tried it could possibly endure the intense agony it is capable of producing'.[15]

Electrical invention could be a cut-throat business, clearly. Saxton and Clarke sent furious letters to the scientific press defending their own claims to novelty and insulting one another. Saxton spluttered that the 'machine which Mr Clarke calls his invention, differs from mine only in a slight variation in the situation of its parts, and is in no respect superior to it'.[16] Clarke lashed back that quite a lot might depend on that 'slight variation'.[17] Competition aside, though, by the end of the 1830s, Clarke and Sturgeon, at least, were fellow members of the London Electrical Society – and Saxton was back in Philadelphia. Sturgeon had founded the society to further the cause of practical electricians like himself. There, 'rarely a month passes, without some important new fact being announced, or some new apparatus exhibited'.[18] The first meetings were at the Adelaide Gallery (where Sturgeon was a lecturer) before moving to the Royal Polytechnic Institution on Regent Street. They provided opportunities for members to show off their ingenuity. On 4 November 1838, for example, Clarke demonstrated the latest version of his magneto-electric machine that could generate 'large and brilliant sparks, sufficiently so that a person can read small print by the light it produces'.[19] They were anticipating a future in which electricity would be 'usurping, at no very distant period, the place of steam'.[20]

On 17 July 1838, a Mr Coombs demonstrated a model of Thomas Davenport's 'electro-magnetic locomotive machine' to the society. Davenport, an American blacksmith, had patented his rotary electromagnetic engine in the United States and his agents were attempting to drum up some European interest. Davenport had come up with the plan for his device after seeing an electromagnet designed by Joseph Henry.[21] Henry, then the professor of natural philosophy at Princeton, had taken William Sturgeon's electromagnet and made it a far more powerful instrument. By carefully insulating the wires (with silk from his wife's underwear, according to one story) and dividing them into a number of coils, each attached to a different battery, he could make magnets capable of lifting several hundred pounds. The electromagnet he built for his patron Benjamin Silliman at Yale College could lift 700 pounds.[22] When Davenport saw one of Henry's magnets at work he was impressed by its power. He bought one, took it apart and worked out how to put it together again. With this practical knowledge of electromagnets, he set about making a rotating electromagnetic motor that worked by magnetic attraction and repulsion as electromagnets were consecutively switched on and off. It was not the first such motor, though, as William Sturgeon pointed out angrily following Coombs' performance before the London Electrical Society. Both Sturgeon and the electrician William Ritchie had built similar devices before Davenport.[23]

Electromagnets and electromagnetic engines depended on a reliable source of electricity to power them. Since their invention by

Alessandro Volta in 1800, voltaic, or galvanic, piles were the most common sources of current. At the Royal Institution, Sir Humphry Davy boasted a battery made from as many as 2,000 cells.[24] But even Davy's great battery suffered from the problem that plagued all batteries. It was difficult to manage and the current tended to peter out quite quickly. John Frederic Daniell, the professor of chemistry at King's College London, came up with one solution in 1836 with the invention of a new and constant battery.[25] A few years later, William Robert Grove developed his own nitric acid cell, which was far more powerful than Daniell's battery.[26] Making better and more efficient batteries was a constant preoccupation of the London Electrical Society's practical electricians. They competed (sometimes bitterly) with each other for the best performance, showing off their batteries' ability to generate impressive sparks or a copious decomposition of water. John Shillibeer, for example, bragged about how his combination 'requires but little food, and with that will perform a good honest day's work'.[27] Economy mattered if electricity was going to replace steam. One of the reasons Grove's nitric acid battery was so highly prized was its economic performance.

Grove's battery was seriously tested in St Petersburg during the autumn of 1839. As the natural philosopher Moritz Hermann von Jacobi wrote to Michael Faraday: 'I made, as you may perhaps have learned by the gazettes, the first experiments in navigation on the Neva, with a ten-oared shallop furnished with paddle-wheels, which were put in motion by an electro-magnetic machine.' The machine could 'produce the force of one horse (steam-engine estimation),' he said.[28] Performing at one of Faraday's Friday Evening Discourses at the Royal Institution, Grove himself was quick to point out, once the letter had been published in the *Philosophical Magazine*, 'that the batteries employed were on Mr Grove's construction'.[29] A couple of years later, one of Grove's Swansea friends, the engineer Benjamin Hill, built his own electromagnetic engine and put it through its paces in front of the London Electrical Society. The engine, he told them, 'with four small pairs of Grove's batteries, revolves with sufficient power to turn small articles'.[30] In 1848, when the British Association for the Advancement of Science held its annual meeting in Swansea, another of Grove's local acquaintances, the industrialist John Dillwyn Llewellyn, invited visitors to his extensive estates at Penllergaer, just outside Swansea, to see a 'boat, impelled by the Electrical Current ... at work on one of the Lakes'.[31] It was powered, of course, by Grove's batteries.

Boats were not the only things that could run with Grove's batteries. Grove had also been experimenting on the possibilities of electric light. He tried using sparks and 'could occasionally keep up a steady voltaic light in attenuated nitrogen for four or five hours'. Platinum turned out to be a better prospect, though. A tightly packed coil of platinum wire produced a light 'too intense for the naked eye to support, and amply sufficient for the miner to work by'.[32] At one of his opulent scientific soirées, the wine merchant John Peter Gassiot showed off one of Grove's electric lights so bright that 'through the window, it penetrated the outer darkness, shooting over the lawn; but now softened into the sweetest moonlight, and yet clothing the shrubs and turf with intense green'.[33] In 1848, the engineer William Edwards Staite put one of his electric lights on Nelson's Column in Trafalgar Square, where it 'produced the same sort of illumination as the sunlight through atoms of dust'.[34] The following year, a new ballet, *Electra*, was performed at Her Majesty's Theatre, with Staite's light playing the starring role.[35] With patents secured, Staite was one of a number of electricians who were hoping to make a tidy profit from illuminating the future.[36]

Inventors of electrical locomotives were heading to the patent office as well. In November 1839, William Henry Taylor took out a

The ballet *Electra,* with electric illumination centre stage.
Illustrated London News, 1848

patent for an electromagnetic engine.[37] The *Inventor's Advocate* editorialised that 'enough has already been done to show that the power exists, and that it is capable of being applied. Its extent seems to be unlimited, while the cost of its production is comparatively trifling.'[38] Hopeful electrical inventors were heading for the exhibition halls too. Taylor put his engine on show at the Colosseum near Regent's Park. It was a marvel: 'Nothing can be more continuously regular or beautiful than the motion imparted to the wheel – the agent (unlike fire, water, or steam) invisible, yet its effects palpable to the senses.'[39] The Glaswegian instrument maker Robert Davidson had his electrical locomotive – the *Galvani* – on show at the Egyptian Hall in London's Piccadilly. *The Railway Times* were converts, marvelling that the engine's 'simplicity, economy, safety and compactness, render it a far more valuable motive power than steam on railways and in navigation.'[40] It really did look as if the age of steam was on the brink of being replaced by the age of electricity.

However, the difference between fact and fancy in these electrical speculations was difficult to discern. The men who invented these batteries, engines and schemes for electrical illumination had their feet firmly grounded in the realities of experiment. They were proud of their hard-won practical skills, gained through an intimate knowledge of the magnets, the coils, the acids and metal plates that made up their electrical universe. They were proud of being the men who really knew how to put electricity to work. But even their most pragmatic pronouncements about what their contrivances could do slipped easily into speculation. The urge to speculate (in both senses) was in their veins. It had to be, in fact – men of moderate means did not spend money on patents, or even on the raw materials of electrical invention, unless they were reasonably confident that they would get a return for their investment. In effect, they were gambling on the future. As the *Inventor's Advocate* put it: 'There is no object, indeed, to which the attention of the scientific world is now directed, that appears so fraught with important consequences to the interests of mankind, and the development of scientific truth, as the science of electro-magnetism.'[41] Exhibits of electromagnetic engines and electric lights at places like the Adelaide Gallery offered glimpses into the future that electricity offered.

But there had to be some sober calculation mixed in with all this speculation. One early member of the London Electrical Society was the brewer's son from Manchester, James Prescott Joule. As early as the beginning of January 1838, he sent the first of many letters to William Sturgeon, which were duly published in the *Annals of Electricity*, not just describing his electromagnetic engines but describing his efforts to calculate their efficiency and economy. Joule was sure, just like his fellow electricians, that the future was electrical. 'I can scarcely doubt that electro-magnetism will eventually be substituted for steam in propelling machinery,' he argued. 'If the power of the engine is in proportion to the attractive force of its magnets, and if the attractive force is as the squares of the electrical force, the economic effect will be in the direct ration of the quantity of electricity, and the cost of working the engine may be reduced ad infinitum.'[42] It was the economics that mattered, and that was why Joule devoted so much attention to comparing magnets and experimenting with new ways of making them. He compared electromagnets with solid cores with ones made out of closely packed bundles of wire. He suggested that the wire in the coils could be square rather than round to fit more snugly around the iron core.

Fundamentally, Joule was interested in duty. That was the term engineers used to talk about the economic efficiency of their engines. It wasn't just working that mattered but working economically – and that needed careful calculation. What Joule wanted to know was how much of the zinc in the voltaic battery providing electricity to run an electromagnetic engine would be consumed to do a fixed amount of work. That zinc, ultimately, was the fuel that was used up in running the engine, just as coal was the fuel used up in running a steam engine. His experiments told him that 'the maximum available duty of an electromagnetic engine worked by a Daniell's battery' was '80 lbs raised a foot high for each grain of zinc consumed.'[43] Joule was not the only one making these kinds of calculations. William Robert Grove reckoned it would cost £1, sixteen shillings and 10½ pence to keep an electromagnetic engine working at one horsepower with one of his nitric acid batteries for 24 hours. In which case, he concluded, it was 'evident that the expense of electro-magnetic machines far exceeds that of steam: indeed, it could hardly be expected to be otherwise, as with the one we use for fuel manufactured materials, in the production of which coals, labour, &c., have been expended;

in the other, coals and water are used directly'.[44] This was simple economics.

Grove was not hopelessly pessimistic, though – and neither was Joule really. A decade later, Grove suggested that 'if, instead of using zinc and acids, which are manufactured, and comparatively expensive materials, for the production of electricity, we could realise the electricity developed by the combustion in atmospheric air, of common coal, wood, fat, or other raw material, we should have at once a fair prospect of the commercial application of electricity'.[45] But what would happen when what Grove called 'that granary of force, the coal fields', was exhausted? Everyone knew that Britain's imperial and industrial might rested on coal. What would the Empire do without it? That was when electricity might really come into its own. His own theory of the correlation of physical forces was the key, and 'there is every encouragement derivable from the knowledge that we can at will produce heat by the expenditure of other forces'. He pointed to the 'remarkable applications of the convertibility of force' that had 'been recently attained by the experiments of Mr Wilde', who had found that 'by conveying electricity from the coils of a magneto-electric machine to an electro-magnet, a considerable increase of electrical power may be attained', and repeating the process with more machines, even more power was produced. As long as there was a sufficient source of power to run the machines, 'there hardly seems a limit to the extent to which mechanical may be converted into electrical force'.[46]

Not everyone thought the electrical future was quite so assured, though. 'I do not think there is any means of preserving our manufacturing supremacy after our coal fields have been exhausted,' was John Tyndall's response to a question from the political economist William Stanley Jevons. 'We are sure for example to be beaten by America,' he told him.[47] A week or so later, Tyndall – who was

Michael Faraday's successor at the Royal Institution – drew Jevons' attention to an article of his from the *Fortnightly Review* on 'The Constitution of the Universe'. 'I see no prospect of any substitute being found for coal as a source of motive power,' said Tyndall. 'We have, it is true, our winds and streams and tides; and we have the beams of the Sun as a source of the power derivable from trees and animals. But these are common to all the world. We cannot make head against a nation which, in addition to those sources of power, possesses the power of coal.' And what about electricity? 'I agree with you in thinking the talk about electricity as a source of practical power fit to replace coal as so much nonsense,' was what he had to say on that score.[48] Underlying all this was the recognition that Britain and its empire lived or died by its sources of power – and that its future would depend on energy too.

Alternative energies

Tyndall's cavalier dismissal of electricity was unusual – though his pooh-poohing of the idea is further evidence of just how prevalent such electrical boosterism was. There were also other potential competitors to coal. Different kinds of internal combustion engines had been proposed as alternatives to steam since the 1830s, though being usually powered by gas (produced from coal), they were just as dependent as steam engines on a plentiful supply of coal. But by the 1870s, engineers were experimenting with different fuels. Both in Europe and North America, engines fuelled by oil were being made. In 1872, the English-born and Boston-based engineer George Brayton patented his Brayton's Ready Motor that could be fuelled by oil or paraffin. The Brayton Ready Motor was on show at the Philadelphia Centennial Exhibition in 1876. It impressed the *Scientific American*'s editor, Orson Desaix Munn, who pronounced that 'the distinguishing features of this engine are that it can be started in a very short time, that it is economical in its consumption of fuel, and that, owing to the constant maintenance of combustion, it is claimed, the danger of explosion or the hydrocarbon vapor is so greatly reduced as to be practically obviated. It could be stopped quickly, too, they noted: 'Another important feature of the motor is that the consumption of fuel ceases the instant the engine is stopped, the stoppage being effected by simply shutting off the supply of air.'[49]

It was not long before someone used Brayton's motor to generate electricity, though, the *Scientific American* announcing that it generated 'a stream of electricity or electric light, having an illuminating power equal to that of 234 of the lamps mentioned, showing that three times more light may be produced from a given quantity of oil, if its energy is converted first into mechanical power and then into electricity, than if the oil is directly burned in a lamp.'[50] This spelled efficiency and versatility – characteristics that really mattered in the highly competitive world of late nineteenth-century power creation. That versatility was at the fore when, a couple of years later, the Irish engineer John Philip Holland used a Brayton engine to run the *Holland 1* – his experimental self-propelled submarine. Steam simply was not viable for underwater propulsion. His submarine's possibilities even attracted attention from Irish revolutionaries – Holland was commissioned by the Fenian Brotherhood to build the *Fenian Ram* submarine as a weapon to attack the Royal Navy. By the 1880s, a number of engineers had developed vehicles of various kinds that were powered by internal combustion engines. In 1884, Edward Butler took out a provisional patent for a petroleum-powered tricycle and by 1888 was manufacturing the Butler Petrol Cycle.[51]

Petrol did not seem as fantastically promising as electricity, though. When Edward Bulwer-Lytton published *The Coming Race* in 1871, the future it played with was entirely electrical. When its narrator woke up at the bottom of the chasm into which he had fallen, it was to a subterranean world dominated by electricity. The ruling race – the Vril-ya (tall, angelic and winged beings) – did everything with electricity, or 'vril', as they called it. They communicated telepathically by means of electricity; they controlled the weather and their environment by means of electricity; they illuminated the underworld they inhabited with electric power; they travelled in vehicles driven by electricity; they wielded vril staves to strike at their enemies with lightning. The Vril-ya had discovered that electricity was 'capable of being raised and disciplined into the mightiest agency over all forms of matter, animate or inanimate'. It can 'destroy like the flash of lightning; yet, differently applied, it can replenish or invigorate life, heal, and preserve'.[52] The Vril-ya had, quite literally, carved out their place in the world with electricity. The Vril-ya might occupy the Victorian underground, but they lived in the Victorian future, too. They made plain what the electrical future promised – it promised control and apparently limitless power, as long as it was subjected to the right kind of discipline. That was what the Victorian culture of accuracy and precision was meant to provide.

It was so clear to most people that the future would run on electricity that bits and pieces of electrical gadgetry might be casually dropped into scientific romances to give it all a touch of realism. When Jules Verne took Axel and his Uncle Lidenbrock on their *Journey to the Centre of the Earth* in 1864, he made sure, for example, that they were supplied with a lamp powered by a Ruhmkorff coil. 'This ingenious use of electricity,' he said, 'would enable us to go on for a long time by creating an artificial daylight, even in the midst of the most inflammable gases.'[53] When Captain Nemo went *Twenty Thousand Leagues under the Sea* in 1870, he was similarly supplied. Electricity was the heart and soul of Nemo's submarine. As he explained to his unwilling guests: 'There is a powerful agent, obedient, rapid, easy, which conforms to every use, and reigns supreme on board my vessel. Everything is done by means of it. It lights, warms it, and is the soul of my mechanical apparatus. This agent is electricity.'[54] Nemo's visitors might have gasped in surprise at that revelation, but it would not have surprised his readers. Verne even had Nemo describe how he produced electricity directly from seawater to fuel the *Nautilus*.

Verne saw some of the electrical gadgetry that powered the *Nautilus* on show at the Paris *Exposition Universelle* in 1867. The *Exposition Internationale d'Électricité* held there in 1881 inspired even more electrical romancing. Albert Robida's Paris of the future was positively overflowing with the stuff.[55] In 1950s Paris (the romance was set in 1952), Robida had his characters travelling through the air in airships guided by electric lighthouses before landing and continuing in underground electrical trains. Their apartments were powered by electricity and they even ate food cooked by electricity. Paris was defended from attack by batteries of electrical weaponry. Just as Robida's gentle satire of the electrical future was inspired by electrical exhibitionism, *Punch* magazine was inspired by the International Electrical Exhibition of 1882. In 'The Coming Force' (a clear allusion to Bulwer-Lytton's romance a decade or so earlier), Mr Punch was pictured dreaming about the shape of electrical things to come. A troupe of electric sprites surrounding a chariot bearing a dynamo and electric light were driving off the chimney sweeps and gas lighters of the filthy present. There was a man on an electric tricycle and displays of fruit and veg electrically cultivated. Bringing up the rear was a couple of oxen, transported frozen from the Antipodes and revived (for the slaughter, presumably) by electricity.[56] Here was the full panoply of electric possibility on show, and the satire only worked because it seemed to many so plausible.

Books like A.R. Hope's *Electricity and its Wonders* (1881) and J.M. Munro's *The Romance of Electricity* (1893) evoked futures powered by electricity just as far-fetched as anything Mr Punch could dream about. Munro, published by the Religious Tract Society and given out as rewards for faithful Sunday school attendance, speculated about 'passing ships or flying-machines telegraphing to each other on the voyage'. He thought that 'electric light will, of course, become very general', and speculated about the recent experiments of Dr Hertz of Carlsruhe, showing that light itself is 'merely an electro magnetic disturbance of the luminiferous ether'. If that were the case, then 'we may discover a way of directly producing these electromagnetic waves; in short, of manufacturing light'. If that happened, then the 'dynamo, such as we know it, may eventually become as extinct as the dodo'. 'Can we doubt', he asked, 'that there is a marvellous future before this wonderful agent', and that the 'day is coming when the world will be endowed, not only with sensory, but with motor nerves, and when this higher stage of material organization has been reached, we may expect the political and social state of human life to be exalted with it'.[57]

In the year 2000 imagined in John Jacob Astor's *A Journey in Other Worlds*, electricity was ubiquitous. Astor, of course, had substantial financial interests in securing an electrical future like the one he imagined in his scientific romance. Among other financial concerns, he was one of the directors of the Cataract Construction Company behind the hugely ambitious plans to produce electricity from the Niagara Falls. A few years later, Nikola Tesla would attempt – unsuccessfully – to interest Astor in financing his experiments in wireless power transmission at Wardenclyffe. In Astor's electrically fuelled future, 'electricity in its varied forms does all work, having superseded animal and manual labour in everything, and man has only to direct'. The power of wind and water was captured to generate electricity; the 'electrical energy of every thunderstorm is also captured and condensed in our capacious storage batteries'; the 'windmill and dynamo thus utilize bleak mountain-tops that, till their discovery, seemed to be but indifferent successes in Dame Nature's domain'. All of this electricity worked to 'run our electric ships and water-spiders, railways, and stationary and portable motors, for heating the cables laid along the bottom of our canals to prevent their freezing in winter, and for almost every conceivable purpose'. Everyone had a windmill on their roof.[58]

By the 1890s, it really was starting to look as if a future fuelled by electricity was getting closer than ever. At successive international exhibitions in London, Paris, Vienna and Chicago, electricity seemed more visible and more spectacular. The machines were bigger and more powerful. There were more and brighter electric lights. The Columbian Exhibition in 1893 was widely hailed as 'a magnificent triumph of the age of Electricity'.[59] It looked as if the future had already arrived on the shores of Lake Michigan. Magazines like *Cassell's*, *Pearson's* or *The Strand* were full of scientific romances that more often than not simply took the electrical future as a given. The same magazines were also full of essays offering sober – and sometimes not so sober – accounts of just what the electrical future would offer. In 1888, the popular magazine *Answers* set its readers the task of predicting the world in 1988. Electricity featured large in almost all the entries. The winner predicted 'electric light used in every house' and 'trains, cars and buses propelled by electricity'. Another entry projected that 'locomotives will be entirely superseded by electric flying machines' and 'all sea voyages will be made by submarine vessels propelled by electricity'. Another one noted that the 'machine now building intended to voyage to the moon is close on completion'.[60]

Beneath the froth of speculation was the hard-headed business of making the electric future economic. William Robert Grove, in his 1866 address to the BAAS, had pointed to Henry Wilde's improvements in magneto-electric machines as a critical step forward. Wilde had been apprenticed to a Manchester engineer before establishing his own business as a telegraph and lightning conductor. In 1862, an alphabetic telegraph he had invented was on show at that year's international exhibition in London. He made his machines substantially more powerful by replacing the permanent magnets used in earlier versions with electromagnets. Like many inventors, Wilde was given to spectacle and understood that showing off was an important part of the business of invention. He showed off the full extent of his machine's power by using it to melt whole bars of iron, for example. As he explained in a communication to the Royal Society, when 'the alternating waves from the magneto-electric machine were transmitted through a piece of No. 20 iron wire, 0.04 of an inch in diameter, a length of 3 inches of this wire was raised to a red heat'. However, with his improvements, when 'the alternating waves from this electromagnetic machine were transmitted through the same sized iron wire as used in the preceding experiment, 8 inches of it were melted, and a length of 24 inches was raised to a red heat'. With an even bigger machine, he could produce a current 'so enormous as to melt pieces of cylindrical iron rod 15 inches in length, and fully one quarter of an inch in diameter'.[61] John Tyndall told the Brethren of Trinity House that Wilde's machine 'far transcends in power all other apparatus of the kind'.[62]

Barely a month after Wilde's communication had been read before the Royal Society, William Siemens and Charles Wheatstone made their own communications, both offering very similar ways of increasing the power of the magneto-electric machine. Siemens' method had actually been developed by his brother, Werner von Siemens, head of the Berlin-based Siemens & Halske electrical engineering company. The Siemens brothers, Carl, Werner and Wilhelm, were by the 1860s the dominant powers in the German electrical industry. More than that, they had made their reach international by the simple expedient of settling in different countries. Carl had settled in St Petersburg and established a local branch of Siemens & Halske there in 1853. Wilhelm Siemens – his name conveniently anglicised as William – was the London representative of the firm. He was well placed in English electrical circles and had even been elected a fellow of the Royal Society in 1862. The Siemens Telegraph Works in Charlton, south-east London, was a powerhouse of electrical innovation, churning out precision apparatus for the telegraph industry. The Siemens invention, William proudly told the Royal Society, 'provides a simple means of producing very powerful electrical effects.'[63]

Charles Wheatstone in turn boasted of the 'more energetic currents' his machine could generate.[64] Wheatstone, of course, had been a prominent figure in telegraphy since taking out the first British patent for an electric telegraph along with William Fothergill Cooke 30 years earlier. Wheatstone was a member of the BAAS committee on electrical standards and another evangelist for accuracy and precision. As it turned out, they had all been overtaken by another electrical engineer, Samuel Alfred Varley, who had taken out a patent for a self-exciting dynamo at the end of 1866. Varley was another member of a telegraph dynasty. His brother, Cromwell, had played a key role in the investigations following the failure of the first Atlantic Cable, and the patent partnership he had forged with William Thomson had earned him a fortune. The speed with which these innovations were appearing reflects the urgency with which electricians were conducting their experiments – and their hopes for the future fuelled by electricity. It was starting to look as if Grove's predictions to the BAAS would be realised far sooner than he had imagined. A new word was coined – 'dynamo' – to describe this new breed of powerful magneto-electric generators.

The electrical landscape was changing rapidly during the 1870s as electricians across Europe and North America headed for the patent offices with their inventions. At Vienna's International Exhibition in 1873, visitors could wander through a Machine Hall that contained powerful dynamos built by Siemens & Halske along with the usual steam engines. They could see Zénobe Gramme's revolutionary direct current generator. For the inordinately wealthy, the electrical future was there already. In 1871, as alluded to earlier, the arms manufacturer William George Armstrong installed a Siemens dynamo powered by water from a reservoir at his Cragside estate, using the electricity to illuminate the house and grounds. He was not the only one: at Hatfield House, the Tory politician and future prime minister, Lord Salisbury, was busy electrifying. By the end of the 1870s, both Armstrong and Salisbury were using the incandescent light system developed by Joseph Swan. In the United States, Thomas Edison was developing his own system of incandescent lighting, similar to Swan's. Edison had turned his Menlo Park laboratory into a factory for patents. Throughout the 1880s, thanks in part to an enviable ability to court publicity, Edison consolidated his position as the leading electrician in America. The press portrayed him as the 'Wizard of Menlo Park', out to make the world electric. His companies were extending their hold in Europe as well – the young Nikola Tesla was briefly employed by the Société Electrique Edison in Paris before heading for the bright lights of New York.

In London, by the end of the 1880s, the ambitious young electrical engineer Sebastian de Ferranti was building a power station at Deptford that was designed to provide electricity for much of the city. Ferranti had been hired in the first place to electrify the Grosvenor Gallery that had opened in Bond Street a few years earlier. The scheme was so successful that the gallery's patron, the Earl of Crawford, set up the London Electric Supply Corporation to take advantage of Ferranti's genius for electrical generation. Edison might be the wizard, but Ferranti was pictured as a mighty Colossus, bestriding the Thames. The Deptford station, built on a site once owned by the East India Company, was the key to their plan to monopolise the capital's electricity. This was the future writ large 'on a scale of unprecedented magnitude'. The grand ambition was 'for the eventual supply of two millions of lights'.[65] Electricity started flowing out from Ferranti's dynamos to illuminate the houses of the metropolitan elite in 1889. Deptford was a huge investment – and in the first instance it was an investment in the fashionable appeal of

Sebastian de Ferranti straddling the Thames as the Colossus.
Electrical Plant, 1889

electricity. 'Electricity, in a great measure, for purposes of house lighting, must be the rich man's light for a long time to come,' they said.[66]

The promises made for Deptford were a sign that electricity was the coming thing. The Columbian Exposition in Chicago in 1893 was ablaze with electric lights. Electricity powered the enormous Ferris Wheel and the 'beautiful little boats' that shuttled visitors back and forth across the lagoons. On top of the Manufactures Building at the centre of the grounds was a massive electric searchlight visible, supposedly 100 miles away.[67] In the world outside the exhibition, electricity was increasingly ubiquitous. Electric street lights were becoming a common sight in towns and cities across Europe and North America. Electric trams carried passengers between suburbs and city centres. The middle classes were turning their homes electrical in increasing numbers. The electrical engineer Arthur Kennelly boasted that 'the adoption of electrical household appliances is daily becoming more widespread, here adding a utility, and there an ornament, until in the near future we may anticipate a period when its presence in the household will be indispensable.'[68] Electricity was beginning to be used on a large scale to power industry too. In 1889, Britain's Tory prime minister Lord Salisbury – who had already electrified his own estate – even predicted that once 'in the house of the artisan you can turn on power as you now turn on gas', the growth of electric power would reverse the evils of the Industrial Revolution as cities emptied and workers returned to simpler lives and times.[69]

Energetic business

None of this happened by accident – and none of it happened as the result of acts of individual genius either. The business of electrification was a business, and a bloody and brutal one too. By the end of the 1880s, Edison and his companies were locked in a commercial battle with George Westinghouse for control of an increasingly lucrative market in electricity. Edison was committed to developing direct current systems, which could distribute electrical current efficiently at low voltages and over comparatively short distances. This was tried and tested technology. Edison had opened his first direct current power station on Pearl Street in New York in 1882. But European investors were backing alternating current systems, like Ferranti's ambitious Deptford scheme, and Westinghouse was soon backing alternating current too in America. Edison went on the offensive, calling alternating current, which could operate at far higher voltages than direct current and be transmitted over far greater distances, the 'current that kills'. He was soon advocating the use of Westinghouse's system as a means of capital punishment – the process might be called 'westinghousing' the victims, he joked. Despite Edison's best efforts though, alternating current was in the ascendancy by the beginning of the 1890s. It offered economies of scale and long-range transmission that direct current could not match.

Westinghouse's victory in the battle of the systems was complete when his company was awarded the contract to provide the ambitious scheme to generate electricity from the Niagara Falls. Back in 1876, when William Siemens had visited America and the falls he had wondered might 'this colossal power actuate a colossal series of dynamos, whose conducting wires might transmit its activity to places miles away?'[70] The physicist William Thomson also thought that Niagara might be an almighty source of electric power. By the beginning of the 1890s, plans were coming to fruition. The Cataract Construction Company contracted Westinghouse to provide them with ten massive dynamos, each capable of generating 5,000 horsepower. It was 'a gigantic engineering enterprise that has no precedent in the civilised world'.[71] George Forbes, the project's consultant engineer boasted that at Niagara people could 'see a whole new world created'.[72] This really did look to many like the end of coal and steam.

This was power that could 'be sent much more than a hundred miles, and still be more economical than steam, even though coal is cheap there'.[73] Niagara and its powerful generators were 'the nearest obtainable approach to perpetual motion'.[74]

One of the factors behind Westinghouse's success was his purchase of Nikola Tesla's patent for his revolutionary polyphase motor that operated by alternating current in 1888. This was the missing link in Westinghouse's plans, since most existing motors worked by direct current and were cumbersome to use with alternating current systems. In 1888, Tesla was a relatively recent arrival in America, having landed in 1884 to work for Edison, but who soon abandoned his erstwhile employer to establish himself independently. Tesla was a dreamer of fantastical electrical dreams. His reputation made with the success of his polyphase motor, he set out to try and remake the electrical future in his own image.[75] By the beginning of the 1890s, catapulted into the headlines by a series of spectacular lectures in America and Europe, Tesla was the electrical man of the moment. In reality, he had little to do with the grand plans at Niagara, but that did not stop newspapers from describing him as the visionary genius behind it all. He had his own display of his electrical inventions at the Chicago Columbian Exposition. Thomas Commerford Martin told the readers of *Century Magazine* that, thanks to Tesla, when it came to electricity the 'fanciful dreams of yesterday' would soon become 'the magnificent triumphs of tomorrow, and its advance towards domination in the twentieth century is as irresistible as that of steam in the nineteenth'.[76]

Tesla's grand ambition was to develop a system that could send huge quantities of electrical energy pulsating through the ether – enough to power factories and illuminate entire cities. The *Pall Mall Gazette* predicted that if 'Mr Tesla succeeds in making half his discoveries available for daily use, we shall have everything at our

Nikola Tesla's personal exhibit at the Columbian Exposition.
Westinghouse Electric Corporation Photographs, 1893

command that the Vrilya had and shall have gone a long way towards acquiring the amazing forces of the Martians'.[77] Tesla spent much of the 1890s in a desperate quest for money to help realise his ambition. He approached John Jacob Astor but was rebuffed, but eventually he persuaded J.P. Morgan to advance him $150,000. With this, Tesla purchased land at Wardenclyffe, 65 miles from New York, where he started building the apparatus that would allow him to realise his dreams. At its centre was a tower 187 feet tall with a 55-ton metal hemisphere at its apex. The tower would send the electricity generated by a 350-horsepower dynamo hurtling through the atmosphere, where it could be recovered by anyone possessing the right kind of apparatus. 'We are building for the future,' Tesla grandly told the newspapers.[78] Locals told the press about the 'blinding streaks of electricity which seemed to shoot off into the dark on some mysterious errand'.[79]

Wardenclyffe turned out to be no more than pie in the sky, and Tesla's dreams of an electrical future powered by streams of wireless electricity came to nothing. It came to nothing, in part at least, because Tesla refused to learn the most important lesson of Victorian invention — that invention could never be a one-man show. Producing the electrically fuelled world that was starting to emerge by the end of the Victorian era was a collective effort. It depended entirely on the development of new ways of knowing and doing. It depended on the systematic exploitation of the natural resources needed to put electricity to work efficiently and economically. The electrical future depended on copper mined in the Americas and smelted in Swansea in South Wales ('Copperopolis', they called the town). It depended on gutta-percha from the Malay archipelago and cotton from the southern United States to insulate the wires. Committees of sober-minded scientists and engineers, meeting at international exhibitions, worked to establish the electrical standards that underpinned all this. It was also a matter of commerce – and successful electrical entrepreneurs recognised that scientific and commercial standards had to add up to the same thing. As William Thomson, who was keenly aware of the money-making prospects of the electrical future, put it: 'When electrotyping, electric light &c become commercial we may perhaps buy a microfarad or a megafarad of electricity ... if there is a name given it it had better be given to a real purchaseable quantity.'[80]

Away from Tesla's electric dreams, electrification across Europe and America was proceeding rapidly. By the end of the nineteenth century, even relatively small towns were investing in electricity and household electricity was no longer the preserve of the wealthy. People now could – and did – buy purchasable quantities of electricity, delivered into their houses through cables, just as gas was delivered through pipes. In London, as in other cities, electrical supply companies competed fiercely with each other – and with gas companies – to provide electricity for domestic and industrial use. Those international exhibitions where electricians gathered to decide on electrical standards were increasingly dominated by electrical machinery. The first electric tramcar had been put on show in 1882 by Radcliffe Ward at the North Metropolitan Tramways Company in Leytonstone. It took a trip down Union Road 'to the amazement of the inhabitants who, for the first time in their lives, saw a tramcar full of people travelling at the rate of seven or eight miles an hour without any visible motive power'.[81] Just a couple of years later, Thomas Parker was driving around in an electric car, powered by the same kind of powerful accumulator battery that Ward used to run his tramcars. There was plenty of real electrical technology around to provide food for speculation about what the breakthrough might be.

When radioactivity was discovered at the end the century, there was excited speculation that it, too, could become a source of huge power. In February 1896, the French physicist Henri Becquerel had announced to the French Academy of Science that there appeared to be strange and mysterious rays emitted by uranium salts. A few years later, Marie and Pierre Curie identified two new elements – they named them polonium and radium – that appeared to be particularly strong sources of these rays. It soon became clear that these strange rays were coming from inside the atoms of different elements. William Crookes speculated that 'if half a kilogram were in a bottle on that table it would kill us all'. He thought that a single gramme of radium would be 'enough to lift the whole British fleet to the top of Ben Nevis; and I am not quite certain that we could not throw in the French fleet as well'.[82] Just like electricity, radioactivity fired the imagination with the possibility of new sources of power that would transform the future. Like electricity, it offered new ways of thinking about what the possibilities of the future might be, and new ways of speculating about how that future might be fuelled.

The prosaic reality of power at the end of the Victorian era remained steam-driven, of course. There might be electric boats, and cars, and trains, and trams, but most people still travelled by steam. The descendants of Stephenson's *Rocket* still thundered down the railways. It was coal and steam that powered the dynamos that generated the electricity to light late Victorian city streets and houses. Steam technology might not fire the imagination in the way that electricity did, but it was technology that worked. By the end of the nineteenth century, steam engines were highly sophisticated and precision-engineered technological marvels. They were the products of the accumulated scientific and practical expertise of decades. In fact, they were potent examples of the transformative impact of technology. They were built for, and helped sustain, a culture built around technological expertise. Despite (or maybe because of) their ubiquity, steam engines looked less and less like the technology of the future. Nobody thought that the Victorians would get to the Moon by steam. Electricity was the future fuel of choice. It was electricity that fuelled Captain Nemo's submarine explorations. It was electricity that propelled John Jacob Astor's adventurers to Jupiter and beyond. When a pulp fiction author imagined Thomas Edison leading a fleet of spaceships to Mars to take their revenge for the Martian invasion of Earth, it was electricity that fuelled them. There wasn't really any other possible choice of power.

Chapter 6

Surveillance

The first annual dinner of the Institution of Electrical Engineers took place on the evening of 4 November 1889, in the distinctly opulent setting of the Criterion, on London's Piccadilly Circus. Opened sixteen years earlier and designed by the architect Thomas Verity, the Criterion was – and was meant to be – a place for the powerful to rub shoulders in luxury. In its clubbable corners and banqueting rooms, deals were struck and egos stroked. Accordingly, when Robert Cecil, Conservative prime minister and third Marquess of Salisbury, staggered to his feet at the end of an excellent dinner, he was surrounded by the good and the great of electricity. There were 220 men there that night. The institution's president, Sir William Thomson, sat at the head of the table, with the marquess on his right. Around them on the high table sat George Gabriel Stokes, the president of the Royal Society, the Astronomer Royal William Christie, William Henry Preece, the Post Office's chief electrician, as well as professors, industrialists and politicians galore. They were all in celebratory mood. As Salisbury quipped at the end of his speech as he toasted his hosts: 'An institution which dines for the first time may be looked upon as a child that has just been weaned. Well, the simile is not so improper, because it has just taken to its bottle.'[1]

Salisbury's speech was just what his audience wanted to hear. He reminded them just how much electricity mattered for the modern world – one newspaper entitled their report 'Lord Salisbury on Electricity and Civilization'.[2] The telegraph, he pronounced, had for ever changed culture. 'I think the historian of the future when he looks back will recognize that there has been a larger influence upon the destinies of mankind exercised by this strange and fascinating discovery than even in the discovery of the steam engine itself,' he said. It was 'a discovery which operates so immediately on the moral and intellectual nature and action of mankind'. Its working had 'assembled all mankind upon one great plane, where they can see everything that is done, and hear everything that is said, and judge of every policy that is pursued at the very moment those events take place. And you have by the action of the electric telegraph combined together almost at one moment, and acting at one moment upon the agencies which govern mankind, the opinions of the whole of the intelligent world with respect to everything that is passing at that time upon the face of the globe.'[3]

The electricians lapped it up. The speech, as reported in a number of newspapers, was regularly interrupted with enthusiastic cries of 'Hear! Hear!' As a wily and cautious politician, the Marquess of Salisbury knew exactly what to say to tickle his audience's vanity, but he meant what he was saying, nonetheless. He invited them to contemplate 'the existence of those gigantic armies held in leash by the various Governments of the world'. 'What gives those armies their power?' he asked. 'What enables them to exist? By what power is it that one single will can control these vast millions of men and direct their destructive energies at one moment on one point? What is the condition of simultaneous direction and action which alone gives to these vast armies this tremendous power?' His audience knew the answer already: 'It is nothing less than the electric telegraph.'[4] The claim elicited another round of 'Hear! Hear!' – Salisbury was, after all, putting their business right at the heart of empire, where they wanted to be. But this really was what Victorian politicians and Victorian engineers thought the telegraph offered, and Salisbury, whose government was leading the scramble for Africa, understood that very well indeed. It was all about the complex interconnections of power, intelligence and oversight – and it had been from its very beginnings. In all sorts of ways, the telegraph was such a successful piece of imperial technology because that was exactly what it had been designed to be.

When William Fothergill Cooke started jotting down ideas for his great invention in 1836, he was already alert to its wider possibilities. Just as Chartist agitation was starting to spread, Cooke pointed out that by means of his telegraph 'in case of dangerous riots or popular excitement, the earliest intimation thereof should be conveyed to the ear of the Government alone, and a check put to the circulation of unnecessary alarm'. He imagined networks of government agents spread across the country, equipped to take over telegraph stations at need, so that 'the Government would be enabled in case of disturbances to transmit their orders to local authorities, and, if necessary, send troops for their support; while all dangerous excitement of the public might be avoided'.[5] In fact, a clause to this effect was inserted into the Electric Telegraph Company's charter, and duly invoked during the Chartist uprisings of 1848. The telegraph was designed to make the world seem smaller and easier to manage. Looking back at the early days of telegraphy, the electrical engineer Latimer Clark, who had been a telegraph engineer almost from the beginnings of telegraphy, marvelled that 'distance and time have been so changed to our imaginations, that the globe has been practically reduced in magnitude, and there can be no doubt that our conception of its dimensions is entirely different to that held by our fathers.'[6] It was a common – and revealing – sentiment.

Just as revealing were other metaphors. The telegraph, said the *Patent Journal*, was like 'a dutiful child or obedient servant' that had been 'trained to carry our messages through the air by the road we have made for it, and with equal velocity through the earth by a road it makes for itself'. It crossed 'the mighty deep in the shape of an angel of peace, bearing the olive branch to countries formerly our most bitter and inveterate foes'.[7] It was, said the journalist Andrew Wynter, 'a spirit like Ariel to carry our thoughts with the speed of thought to the uttermost ends of the earth'. Wynter indulged in another telling metaphor too. He called the Electric Telegraph Company's head-quarters in Lothbury Square 'the great brain … the nervous system of Britain'.[8] Again, the comparison of the telegraph to the nervous system was a common one. Responding to Salisbury's speech, the president of the Royal Society remarked that now 'the whole earth resembles, in a measure, our own bodies. The electric wires represent the nerves, and messages are conveyed from the most remote regions to the central place of government, just as in our bodies, where sensations are conveyed to the sensorium. And then, again, the orders are issued.'[9] That corporeal metaphor was about control and command too.

One of the stories repeated most often about the telegraph was its role in the arrest of John Tawell for murder in 1845. A former convict who had returned to England from Australia in 1838 a relatively wealthy man, Tawell had installed his mistress, Sarah Hart, in a cottage near Slough, supporting her and their two children with the sum of £1 a week. By 1844, the money had run out, and on 1 January 1845, Tawell took Brunel's Great Western Railway from London's Paddington Station to Slough to murder Sarah by poisoning her with prussic acid. He was spotted leaving her house and at the local railway station as he attempted to flee back to London. Once he was on the train, Tawell must have imagined he was free from further pursuit, but what he did not know was that the telegraph had been extended to Slough along the line of the Great Western. His description was duly telegraphed to Paddington Station ('he is in the garb of a Kwaker' the message said, since the simplified alphabet then used by the telegraph had no 'Q' for 'Quaker') where he was followed by police and duly arrested, tried and executed.[10] The whole affair brought home to the public that, with the telegraph, information could now travel more quickly than mere humans.

Newspapers published anecdotes, apocryphal or otherwise, about criminal plots foiled, fare dodgers prosecuted and eloping couples caught by the telegraph. As one put it: 'Parents of marriageable children, too, may sleep in tranquility – for Gretna-Green marriages will be hard to effect when the electric telegraph becomes general.'[11] The telegraph, everyone understood, made surveillance, by agents of the state or private individuals, easier to manage. On one occasion, when a couple of *Morning Herald* journalists used the telegraph to report news (false, as it turned out) of an uprising in Ireland, the Electric Telegraph Company's director did what 'every true and loyal subject was bound to do', and passed the information on to Home Secretary Sir George Grey.[12] The *Morning Herald* complained loudly that the company had agreed to let them publish the story first and had reneged on their agreement. 'So much for good faith,' they sneered, and found themselves in court for libel.[13] If telegraphs were taken over by the government during the Chartist uprisings, the Chartists had their own ways of retaliating. At one stage in 1848, John Lewis Ricardo, the Electric Telegraph Company's chairman, called for police protection for telegraph operators who had been threatened by Chartist agents.[14]

Telegraphy soon started to develop its own distinctive culture. There was early discussion of what kind of person would be best suited to operating the telegraph. Children, and even women, were suggested since they would be easily disciplined, and the work would be easy and repetitive. This was a common view that was held about women, children and work. The enthusiast for mechanised factories, Andrew Ure, for example, had suggested that 'whenever a process requires peculiar dexterity and steadiness of hand, it is withdrawn as soon as possible from the *cunning* workman, who is prone to irregularities of many kinds, and it is placed in charge of a peculiar mechanism, so self-regulating that a child may superintend it.'[15] Children (and women), it was supposed, were good at repetition, and good at doing what they were told. William Fothergill Cooke suggested people who were profoundly deaf, since they were already adept at using a symbolic language of the kind that would be used by the telegraph. What operators really needed, as it turned out, was speed. They needed to be able to read, record and transmit messages down the line as quickly as possible.

In practice, telegraph operators were usually young men, in their late teens or early twenties. They were the sort of young men who might otherwise have been found behind desks in counting houses or accountants' offices. A newspaper article from January 1845 gives a nice glimpse of their culture. At the stroke of midnight on 31 December 1844, the Paddington telegraph operator sent a message down the line wishing his opposite number at Slough a happy new year. His fellow operator shot back that 'the wish was premature, as the new year had not yet arrived at Slough.'[16] There were stories of games of chess played between bored operators at different stations. This was a boys' culture of practical jokes and coded messages sent up and down the wires. They developed tricks of the trade to make their work easier. They could (presumably on purpose) be infuriating: 'The plan is to make as much fuss as possible, and to insist upon the observance of details with the same pedantic precision as if a request to your wife at Brighton to secure a bed for Smith, who is coming down with you, was to be registered among the archives of the nation.'[17] In the United States, itinerant telegraph operators cadged rides on the railroads from station to station as they worked the lines. It was as a telegraph operator riding the rails that Thomas Edison started his career.

When someone like Latimer Clark began to work as a telegraph engineer during the 1840s, there were no experts at telegraphy. This is one reason for William Fothergill Cooke and Charles Wheatstone's acrimonious dispute over who had really invented the telegraph. It simply was not clear what kind of skills and what kind of knowledge would matter for telegraph work. Charles Vincent Walker, for example, had originally been apprenticed as an engineer. He was an enthusiast for electrical experimentation and an early member of the

Playing chess by telegraph.
Illustrated London News, 1845

London Electrical Society, acting as both secretary and treasurer. During the early 1840s, he edited the *Electrical Magazine*, and it was presumably his electrical know-how that led to his appointment as the South-Eastern Railway Company's electrician in 1845. Clark himself had started his career as a civil engineer and had worked with Robert Stephenson on the Britannia Bridge across the Menai Strait, before being employed by the Electric Telegraph Company. He had been an engineer on the Atlantic Cable. By the time Clark became president of the recently founded Society of Telegraph Engineers in 1875, telegraph engineering was a distinct profession, requiring its own kind of expert knowledge. Indeed, the establishment of the society itself in 1871 was a sign of this, just as the decision to rename it the Institution of Electrical Engineers in 1889 signalled the fact that expertise in electricity had by this time extended beyond telegraphy. By then, domestic British telegraph lines had been nationalised and were run by the Post Office. The army had its own cadres of telegraph engineers as it became clear that rapid and effective communication was a prerequisite of imperial warfare.

As submarine telegraph cables wrapped themselves around the globe in the quarter-century or so following the successful laying of the Atlantic Cable, the industry was increasingly dominated by British telegraph companies. There had been abortive attempts to lay a submarine cable along the Red Sea, with a view to extending the line to India, as early as 1859. In 1870, the British–Indian Telegraph Company succeeded in laying a cable from Suez to Aden on the Red Sea, and onwards to Bombay. In 1871, the British Indian Extension Telegraph Company laid a cable from Madras to Penang in Malaysia, while the British–Australian Telegraph Company put down a cable linking Java and Port Darwin. The China Submarine Telegraph Company laid down a cable connecting Singapore and Hong Kong.

By 1876, Australia and New Zealand had been connected by submarine cable. In the Caribbean, the West India and Panama Telegraph Company was busy laying cables. Telegraph cables were understood to be a vital imperial asset. At the Colonial Conference in 1887, delegates called for a cable linking Canada to Australia. 'If there is to be any practical progress made in consolidating the Colonial Empire, the establishment of such new lines of Imperial communication as I have alluded to, by telegraph and by fast merchant cruisers, is to my mind an absolute necessity,' said one.[18] An all-red line – telegraph communication across the Empire that only crossed British-controlled territory – was a key aim by the end of the century. It was vital that the information that flowed through the cables that tied the Empire together be safe from interception.

Those underwater cables, like the rest of the telegraph's infrastructure, depended entirely on the resources of empire, too. Cables were quite intricate devices. Typically, they consisted of several strands of copper wire, usually arranged around a central core. Each strand was isolated with gutta-percha, and often an additional insulator, such as Chatterton's compound (made of gutta-percha, resin and tar), and wrapped in jute – a fibre made from plants heavily harvested in India throughout the nineteenth century. Cables would be strengthened by being cased in a sheath made of iron wires. In the wake of the failed Atlantic Cable of 1858, these cables were very carefully manufactured. The original cable had failed partially as a result of poor manufacture, with sections made by different producers being of different and variable quality. The solution was to insist on common standards that needed careful policing. The materials from which the cables were made came from all corners of the Empire and beyond. Gutta-percha was made from the sap of a tree native to parts of the Malay archipelago, for example. It had been harvested for various uses by the indigenous population for centuries, but having been introduced to Britain in the 1840s it was soon in use as an insulator for telegraph cables – Charles Vincent Walker was the first to use it in this way – and as submarine telegraphy expanded, it was harvested on a huge scale to meet the requirements of the industry. In fact, the rate of harvesting necessary to sustain the need for cables was so great that by the end of the century there was serious concern that the gutta-percha trees would become extinct.[19] Similarly, the copper and iron needed in vast quantities far outstripped what could be mined in Britain itself. Telegraphy was entirely dependent on a global and imperial economy – an economy that it helped sustain, of course.

The Marquess of Salisbury had been quite clear in his speech to the massed ranks of electrical engineers just how important the telegraph was to the Empire. 'The whole work of all the chanceries in Europe is now practically conducted by the light of that great science, which is not so old as the century in which we live,' he told them. The telegraph had made it possible to practise politics at a distance: 'There is a strange feeling that you have in communicating constantly and frequently day by day with men whose inmost thoughts you know by the telegraph, but whose faces you have never seen.'[20] Journalists had been calling the telegraph 'the nervous system of Britain' since the 1850s. Salisbury's point was that it was the nervous system of the Empire too. Imperial power could be exerted at a distance thanks to electricity. That was one reason, at least, for a Tory prime minister to spend a convivial evening at the Institution of Electrical Engineers – though Salisbury keenly embraced electricity in any case, with the family seat at Hatfield House already electrified. The Empire was coming to depend on the globally distributed expert knowledge that underpinned telegraphy. That expert knowledge, in turn, was entirely dependent on resources that only the Empire could provide.

'The greatest marvel'

After Wheatstone and Cooke secured their telegraph patent in 1837, there was a rush to the patent offices and the exhibition halls. In 1838, Edward Davy put his apparatus on show at Exeter Hall. Around about the same time, William Alexander put his telegraph on show at the Adelaide Gallery. In June 1838, Davy acquired a patent for his telegraph, posing a serious commercial threat to Wheatstone and Cooke. While all these telegraphs worked on similar principles, they featured a variety of different ways of receiving and recording the messages sent through them. Wheatstone and Cooke's original invention, for example, used an array of five needles to spell out letters, although they soon developed simpler and less cumbersome apparatus. Another competitor, Alexander Bain, patented a receiving apparatus that worked by using an electrified needle to produce traces on a ribbon of chemically treated paper. In the United States, the artist Samuel Morse patented his telegraph in 1840. Along with his partner Alfred Vail, he developed a simple code of dots and dashes that would be recorded on a ribbon of paper. Morse code soon became the most common way of sending and receiving telegraphs – and skilled operators could write down the messages received just by listening to the sound of the receiver.

Telegraphy clearly offered plenty of opportunities for hopeful inventors as they jostled to persuade expanding telegraph companies across Europe and North America of the advantages of their different apparatus. In 1855, the Welsh–American inventor, David Edward Hughes, patented a printing telegraph in which messages were typed using a simple keyboard and printed out at the receiving end. At the time, Hughes was a professor of music at St Joseph's College in Kentucky, and the printing telegraph bears a startling resemblance to a piano. Simpler and more effective than previous efforts at a printing telegraph, Hughes' instrument was soon in widespread commercial use. As well as telegraphic apparatus for general use, inventors devised instruments for specific purposes, such as fire alarms. The first patent awarded to Thomas Edison was for a vote-recording apparatus on telegraphic principles. As in the case of Hughes' printing telegraph, telegraph companies and investors were on the lookout for innovations that made telegraphy more efficient – faster and more economic. In 1871, Joseph Stearns developed the first practical system of duplex telegraphy that allowed telegraph messages to be sent down the line simultaneously in both directions. A few years later, Edison developed a quadruplex system that allowed four messages to be sent simultaneously (two in each direction).

Unsurprisingly, with a crowd of inventors jostling to be the first to the patent office, telegraphy was a very contentious business. The ink was barely dry on their patent before Charles Wheatstone and William Fothergill Cooke fell out over their respective contributions to the invention. Even after the careful adjudication by Marc Isambard Brunel and John Frederic Daniell, dividing the kudos between them, Cooke kept on worrying at what he saw as his lack of recognition for the rest of his life. Morse in the United States was similarly prickly. He fell out with the co-creator of Morse code Alfred Vail and did his best to contest the contributions that others, such as Joseph Henry and Leonard Gale, had made to the development of his telegraph apparatus. Morse went through endless rounds of patent litigation as he tried to protect his invention from infringement. This litigiousness was the result, in part, of the sheer uncertainty of what really counted as novelty in this time of rapid innovation. It was also, of course, an indication of the fortunes to be made if an inventor really did succeed in doing something original that gave them an edge over the competition.

Alexander Graham Bell was himself very much a product of this highly competitive inventive culture. From the late 1860s onwards, he had been working on what he called a harmonic telegraph, which he hoped would be able to transmit a number of telegraph messages simultaneously down a single wire by means of vibrations of different frequencies. Having emigrated to Canada with his parents, by the early 1870s he was working professor of vocal physiology at Boston University and continuing the family tradition of teaching the deaf. He continued his experiments with the harmonic telegraph, aiming now to use the apparatus to transmit human speech. In 1875, he and his assistant Thomas Watson succeeded in doing just that, and by early 1876 they were successfully transmitting comprehensible speech. Bell received a patent for his invention. He had already established a company – the Bell Patent Association – to fund his experiments and exploit the invention. Another inventor, Elisha Gray, was also working on ways to transmit multiple messages by means of different frequencies. The rivals were aware of one another's experiments, and the more experienced Gray was unimpressed by Bell's efforts. 'Bell seems to devote all of his energy to the concept of vocal telegraphy,' he told his patent agent. 'Certainly, this is of certain scientific interest but is of no commercial interest at the present time.'[21] He was not the only one to think so.

Gray, as he explained to his patent agent, did not see a future for telephony. What was wanted – and what his investigations of sending messages simultaneously at different frequencies aimed to deliver – was more efficient telegraphy. By the 1870s, the sheer volume of telegraph traffic was both enormous and increasing. Anyone who could develop a system for increasing the system's capacity without needing to invest in expensive new lines was sitting on a gold mine. It was not as clear what use vocal telegraphy might have. It was already the case that information could be transmitted by Morse code, for example, much more quickly and efficiently than it might be communicated with the human voice. In telegraphy, speed of transmission and volume of traffic were all that mattered, and the telephone had nothing to contribute to solving those problems. Bell persevered, however, and entered his telephone apparatus as an exhibit at the 1876 Centennial Exhibition held in Philadelphia to celebrate the centenary of the American Revolution. The unprepossessing apparatus attracted little attention at first, but it did eventually come to the attention of the two judges, Joseph Henry and William Thomson, who declared it 'the greatest marvel ever achieved in electrical science' and awarded it one of the exhibition's coveted medals. Before long, Bell was experimenting over long distances. 'Telephones were placed at either end of a telegraph line owned by the Walworth Manufacturing Company, extending from their office in Boston to their factory in Cambridgeport, a distance of about two miles,' and 'articulate conversation then took place along the wire.'[22]

Following up on his triumph at the Centennial Exhibition, Bell continued to exhibit his invention. He understood as well as any inventor that showmanship was as important as experimental expertise in making his instrument part of the future. In his first performance at the Lyceum Hall in Salem, Massachusetts, on 12 February 1877, Bell used the telephone to communicate with his assistant Thomas Watson far away in Boston. Morse code was played down the line in musical notes and 'the audience burst into wild applause'. A 'telephonic organ' in Boston played 'Auld Lang Syne' and 'Yankee Doodle' to the Salem audience and the songs were 'readily heard throughout the hall and heartily recognised'.[23] His performances were soon being copied. In London, *The Times* reported, 'wires have been laid from the Queen's Theatre to the Canterbury-hall, on the south side of the river, and the proprietors of these places have arranged for a telephonic concert, in which the audience in one house will hear the music played at the other'.[24] Bell was at the British Association for the Advancement of Science's annual meeting in Plymouth that summer and lectured to a packed theatre. This time it was 'God Save the Queen' played through the telephone.[25] The following January, 'Professor Alexander Graham Bell, of the Boston University, had the honour of exhibiting the telephone to Her Majesty, Princess Beatrice, and the Duke of Connaught last evening in the Council Room at Osborne.'[26]

While he was performing, Bell was also busy improving his apparatus and further securing his invention. He was not the only one. Thomas Edison had an eye to the telephone's possibilities as well. Both Edison and David Edward Hughes, inventor of the printing telegraph, developed carbon transmitters that were far more effective than Bell's version for transmitting over long distances. Hughes, now based in London, went on to develop a microphone along the same principles. In 1877, Bell, along with his soon to be father-in-law Gardiner Greene Hubbard, established the Bell Telephone Company to develop the commercial possibilities of his invention. They were soon manufacturing telephone equipment to Bell's design, and two years later established the International Bell Telephone Company to market the equipment in Europe. Telegraph companies were also starting to take an interest at last, scenting both competition and lucrative opportunities to expand their activities in new directions. In Britain, telephones fell under the jurisdiction of the Post Office, which already controlled the nationalised domestic telegraph network. Telephones would clearly not be in direct competition with telegraphs – they would never be capable of dealing with the huge volume of traffic – but it was increasingly clear that they offered interesting commercial opportunities to be exploited.

One example of the possible future of the telephone was on display at the Paris *Exposition Internationale d'Électricité* in 1881. A telephone connection to the Paris Opera House enabled visitors to the exhibition to listen to performances at a distance. *Scientific*

American reported that one of the exposition's most popular attractions was 'the nightly demonstrations of the marvelous powers of the Ader telephone, by its transmission of the singing on the stage and the music in the orchestra of the Grand Opera at Paris, to a suite of four rooms reserved for the purpose in one of the galleries of the Palais de l'Industrie'.[27] *The Times*' correspondent was amused to imagine how 'visitors will be able from the Palace of Industry to hear Mdlle. Krauss or Mdlle. Bartet, unless, indeed, they prefer to hear *Hermosa* with one ear and *Phèdre* with the other'. The quality of the sound was not quite perfect yet, they said, but 'all kinds of progress may be expected in the future, and who knows but a few years hence all comfortable houses may be fitted with the means of hearing at the fireside the finest operas, concerts, and tragedies, with the additional advantage of being able to exclude the second-rate pieces with a kind of turncock for music and acting, just as they now have a turncock for water and gas'.[28]

This was clearly one kind of future for the telephone, and scenes like these certainly featured in scientific romances. But the telephone was also finding a niche in everyday business and domestic life during the 1880s. In just a few years, the invention – that 'monument of persevering study and experiment' – had moved from 'being a mere scientific curiosity, universally believed to be of no practical value', to being 'an important factor in the daily business and social life of this and other large cities'.[29] The telephone would result in 'a new organization of society – a state of things in which every individual, however secluded, will have at call every other individual in the community'.[30] As telephone networks spread, it became clear that speed and volume were not the only things to matter for efficient communication. Telephones in the middle-class home were sometimes a source of unexpected disquiet, though. Homes were meant to be domestic sanctuaries, and entry into them strictly policed. There were stories of fraud perpetrated over the telephone, or illicit love affairs conducted under the very noses of protective parents. The 'Telephone girls', who were responsible for connecting callers to each other, were figures of fantasy – but dangerous ones too. They were, after all, privy to conversations that were meant to be private. They would know all about the affairs – personal and financial – of everyone whose calls went through the exchange.[31]

Just like the telegraph, the telephone gave rise to fantasies about a future in which things would be done differently, and social relationships remade through technology. 'The time is close at hand,' argued the *Scientific American*, 'when the scattered members of civilised communities will be as closely united, so far as instant telephonic communication is concerned, as the various members of the body now are by the nervous system.'[32] Just as with the telegraph, that metaphor was a telling one, but it was a metaphor that skated over the resources that went into the telephone network.

In a telephone exchange.
Walter Jerrold, *Electricians and Their Marvels*
(London: S. W. Partridge & Co., 1897)

To the urban middle class for whom the telephone was rapidly becoming just another piece of domestic furniture, using it might be simply a matter of picking it up and making a connection. In reality, sustaining that vision of effortless communication was a great deal of work, of course. The telephone girls were a new kind of worker. Maintaining telephone networks – just as maintaining telegraph networks – needed the cultivation of new kinds of expertise. Telephones themselves were sensitive and sophisticated pieces of apparatus that required highly skilled electricians for their manufacture and maintenance. Pundits admiring the telephone treated its growing popularity as further evidence that there was 'nothing more characteristic of the present age than the avidity with which it seizes upon and puts to practical use the discoveries of science and the infinite marvels of invention'.[33] They usually left out the sheer labour that underpinned the science and invention beneath it all.

Future vision

Bell's telephone offered an inspiration to other hopeful inventors as well. This became clear – if clarity was needed – less than a year after the instrument's successful performance at the Centennial Exhibition, and while Bell was still busily showing it off on both sides of the Atlantic – when the *New York Sun* announced rumours of yet another startling invention. Their correspondent had heard that an 'eminent scientist of this city, whose name is withheld for the present, is said to be on the point of publishing a series of important discoveries, and exhibiting an instrument invented by him, by means of which objects or persons standing or moving in any part of the world may be instantaneously seen anywhere and by anybody'. This astonishing instrument would 'supersede in a very short time the ordinary methods of telegraphic and telephonic communication'. Merchants could sell goods from far away and criminals would be foiled however far they fled. Scholars, wherever they were, would be 'enabled to consult in their own rooms any rare and valuable work or manuscript in the British Museum, Louvre, or Vatican, by simply requesting the librarians to place the book, opened at the desired page, into this marvellous apparatus'. With it, people would be able 'not only to actually to converse with each other ... but also able to look into each others' eyes, and watch their every mien, expression, gesture, and motion'.[34] This was a machine that would do to human eyes what the telephone had already done to the voice.

The newspaper was a little vague as to the invention's details, or indeed the identity of the inventor. Some people thought it was Edison, who had recently announced the invention of what he called a 'telephonoscope' – though that turned out to be something completely different. A couple of years later, *The Times*, along with a number of other electrical journals, announced that 'M. Senlecq, of Ardres, has recently submitted to the examination of MM. du Moncel and Hallez d'Arros, a plan of an apparatus intended to reproduce telegraphically at a distance the image obtained in the camera obscura'. The invention was 'based on the property possessed by selenium of offering a variable and very sensitive electrical resistance according to the different graduations of light'. It consisted of 'an ordinary camera obscura containing at the focus an unpolished glass and any system of autographic telegraphic transmission; the tracing point of the transmitter intended to traverse the surface of the unpolished glass will be formed of a small piece of selenium held by two springs acting as pincers, insulated and connected, one with a pile, the other with the line'.[35] The instrument now also had a new name – the telectroscope. In due course, the *Electrician* published a more detailed account, remarking that Senlecq's telectroscope had 'everywhere occupied the attention of prominent electricians who have striven to improve on it'.

One of those prominent electricians was George R. Carey, who declared in *Scientific American* a few months later that the 'art of transmitting images by means of electric currents is now in about the same state of advancement that the art of transmitting speech by telephone had attained in 1876'. It was a bold claim, and one that underlined the confidence that ambitious electrical men felt in their collective capacities. After all, Bell had a working telephone by then. Carey, a professional surveyor from Boston, offered two different prototypes of the telectroscope, or 'instruments for transmitting and recording at long distances, permanently or otherwise, by means of electricity'.[36] In one arrangement, an image was projected by a camera lens onto a disc containing an array of selenium pieces, each part of an electric circuit, so that when that individual piece of selenium was exposed to light, a current would flow. If the disc at the receiving end was configured in the same way as the transmitter and chemically treated paper placed on it, it would reproduce the original image viewed by the camera. In another arrangement, a selenium point moved by clockwork over a glass plate inside the camera, the idea being that the varying intensity of light shining on different portions of the plate would cause variations in the electric current which would be reproduced at the receiving end.

The common factor in most of the different proposals for telectroscopes, seeing telegraphs and electric telescopes that proliferated around this period in the five years since Bell's exhibition of his telephone at the Philadelphia exhibition was the recently discovered light sensitivity of selenium. The element's sensitivity to light had been noted in 1873 by the telegraph engineer Willoughby Smith who had been experimenting with bars of the metal to determine whether its high resistance would make it useful for underwater telegraphy. During his experiments, Smith had noticed that the bars' electrical resistance varied with the intensity of light to which they were exposed: the higher the intensity of the light, the lower the resistance.[37] Echoing Carey's confidence, the telegraph engineers John Perry and William Ayrton suggested, in their own communication to *Nature* about the prospects of electrical seeing at a distance, that the 'complete means for seeing by telegraphy have been known for some time by scientific men'.[38] Their scheme consisted of a transmitter composed of small squares of selenium on which the image to be sent was projected by a lens. Each square of selenium was connected to a wire leading to the receiver where the image was reconstituted using different methods of electrically opening and shutting apertures through which light was shining.

In light of the rapidity with which Bell's telephone had been transformed from scientific curiosity to sound commercial technology, it was clear that the telectroscope would be the next triumph of applied electrical expertise. It was soon a common feature of speculation about the future. In Albert Robida's *Le Vingtième Siècle* (*The Twentieth Century*) in 1883, telectroscopes – or 'telephonoscopes', as he called them – were everywhere. Set in 1956, the novel described how 'the Universal Theatrical Telephonoscope Company now boasts six hundred thousand subscribers in all parts of the world'. The telephonoscope was a 'simple crystal screen, flush with the wall or set up as a mirror above the fireplace', and the viewer 'simply sits in front of the screen, chooses his theatre, establishes the communication, and the show begins at once'.[39] In an intriguing inversion, a cartoon in *Punch* in 1879, drawn by George du Maurier, was cited by Ayrton and Perry as the inspiration for their proposed telectroscope. The cartoon portrayed a cosy domestic scene with a Victorian paterfamilias and his wife sitting by the fire. On a screen over the mantelpiece, they are watching and talking to their children who are playing tennis in Ceylon – a nice reminder of the imperial reach and purpose of these kinds of technologies.

The telephonoscope.
Punch

Despite the persistent optimism, the telectroscope went nowhere for most of the rest of the century. Then, on 24 February 1897, the Polish inventor and village schoolmaster Jan Szczepanik, along with his financial backer Ludwig Kleinberg, applied for a British patent for a 'method and apparatus for reproducing pictures and the like at a distance by means of electricity'. The application described how 'the picture that is to be rendered visible at a distance is broken up into a number of points and the several rays corresponding to these points are combined together again to form a picture by means of two pairs of mirrors placed at the transmitting and receiving stations and oscillating synchronously'. The process depended on sensitivity and the persistence of vision: if 'the conversion of the several points of the picture that is to be rendered visible take place in sufficiently rapid succession, the eye of the observer will receive the impression of the entire picture, and if the picture be reproduced repeatedly in sufficiently rapid succession, the observer will receive the impression of a permanent picture, the subject of which may appear to be at rest or in motion according to the nature of the successive pictures thus "telectroscoped".[40]

The newspapers were soon full of it. The London *Daily Chronicle's* Vienna correspondent announced to the world that a humble Polish village schoolmaster had succeeded where the greatest electricians had failed and would 'introduce his discovery in the course of the next few days to a select circle of scientific men and journalists', it was said.[41] *The New York Times* announced Szczepanik's claim that 'the telectroscope is, if practicable, as it is claimed to be, something besides a mere toy, an instrument of practical importance, and its efficiency seems to be more considerable than could be judged from particulars hitherto published'. The device could 'not only transmit pictures to a great distance, and reproduce them there, but also that it is possible that it will make the present telegraph instruments super-fluous over short distances'. The *Western Mail* marvelled how, thanks to the amazing invention, the 'man of business may sit at home and see how things are progressing in the office', or a 'lady can see various fashionable costumes at her modiste's', or even that an 'absent lover will be afforded an opportunity of gazing at his sweetheart's face'. As they and other newspapers assured their readers, all would soon be revealed: 'The telectroscope will be first exhibited to the public at the Paris Exposition in 1900.'[42]

Mark Twain even published a short story about the invention and its inventor. 'From the "London Times" of 1904' was written as a journalist's dispatch from the near future and revolved around the invention of an amazing new machine called the telectroscope. Twain's imagined future correspondent described how the device was 'connected with the telephonic systems of the whole world ... and the daily doings of the globe made visible to everybody, and audibly discussable, too, by witnesses separated by any number of leagues.'[43]

Twain's story revolved around a quarrel, a presumed murder and the discovery in the nick of time – thanks to the telectroscope – that the presumed victim was not dead at all, but on the other side of the world enjoying the spectacle of the emperor of China's coronation. Szczepanik himself was one of the protagonists – the presumed murder victim, in fact – and the story started with a quarrel between him and his presumed murderer, Lieutenant Clayton. The quarrel, of course, was about the telectroscope, with Clayton expressing his view that 'the day will never come when it will do a farthing's worth of real service to any human being'.[44] Just like du Maurier's cartoon in *Punch*, the plot rammed home that like the telegraph and telephone that it would soon replace, the telectroscope was going to be another device that brought the world's peripheries closer to its imperial centre.

Szczepanik cut an interesting figure, with a life story that meshed perfectly with the way some people were starting to think about inventors by the end of the nineteenth century – or inventors in the Tesla mould, at least. He was an obscure Polish schoolmaster labouring away in a rural village school who had somehow arrived at the solution to a problem that had confounded the greatest scientists and inventors of the age. Twain had even written about him once before. Just a few months before publishing his short story, *Century Magazine* had published his account of 'The Austrian Edison Keeping School Again'. It was a romantic tale of invention and inventors that some of his readers must have found difficult to distinguish from the fiction they would read a few months later. In 1899, *Pearson's Magazine* published Cleveland Moffett's 'Seeing by Wire', offering another biography of Szczepanik and his invention. Szczepanik, he told his readers, 'as far back as he could remember' had wanted to invent a machine for seeing at a distance. Reading 'stories of Jules Verne, with their dazzle of scientific possibilities' had inspired him to invention, and he had emerged from 'the obscurity of a Galician schoolhouse to a fine eminence with scientists envying him, newsgatherers pursuing him, and capitalists tendering him millions – all at the age of twenty-seven.'

There was, however, one key difference between the telectroscope and its precursors. It did not exist. Despite the hype and the confidence with which successive inventors and reports described the instrument, it was never made. Senlecq and Carey's instruments do not seem to have progressed beyond the drawing board. After their letter in *Nature*, Perry and Ayrton published no more on the matter. Szczepanik's telectroscope did not take the Paris Exhibition of 1900 by storm, after all. In many ways, though, it is the fact that it was never made that makes the telectroscope such a fascinating example of the Victorians' unshakable technological optimism. There were some doubts expressed in the professional press about the solidity of Szczepanik's claims. The *Electrical Engineer*, for example, published a series of notes and editorial commentaries poking fun both at claims and the credulousness of the popular press, suggesting that 'so far Herr Szczepanik has carried out no successful experiment, but has confined his attention to romancing to untechnical reporters.'[45] What the telectroscope really demonstrated was the degree to which confidence in a technologically oriented future was deeply ingrained in the Victorian world view by the end of the century. Not even the *Electrical Engineer* seriously doubted that the telectroscope was viable; they simply doubted that an obscure Polish schoolmaster had achieved it.

Wireless telegraphy

While Jan Szczepanik was busy persuading the press that there really was an instrument called the telectroscope, another hopeful inventor was looking for ways of selling his own revolutionary new device.

For much of the 1890s, Guglielmo Marconi had been working away at the problem of wireless telegraphy. Growing up, Marconi, the son of Giuseppe Marconi, a wealthy Italian landowner, and Annie Jameson, granddaughter of the founder of Irish distillers Jameson & Sons, had been able to indulge his passion for electricity thanks to a succession of private tutors. He had the means, the leisure and the knowledge to make wireless electricity work. Having devoured Heinrich Hertz's accounts of the experiments that had demonstrated the reality of electromagnetic waves in the ether, he worked at extending the distance through which they could be detected. By 1895, Marconi had developed a number of devices that worked by wireless transmission, including an alarm that detected the presence of thunderstorms. But he wanted to go beyond simply demonstrating the possibility of wireless telegraphy. He wanted a practical – and commercial – system. By the end of the summer of that year, he had a system that could transmit signals over two miles and more. A few months later, at the beginning of 1896, he was on his way to London to drum up financial and practical support for his invention.

The British capital was the right place to go looking for backers for many reasons. In the aftermath of James Clerk Maxwell's 1873 *A Treatise on Electricity and Magnetism*, the view that space was filled with an all-pervading electromagnetic ether through which energy moved in waves had become the new orthodoxy for British men of physics. Hertz's detection of such waves in 1888 was simply confirmation of something they already knew, as far as Maxwellian disciples like William Crookes or Oliver Lodge were concerned. Lodge, professor of physics at University College, Liverpool, since 1881 and an accomplished producer of electrical spectacle, had himself been experimenting on wireless transmission. Marconi had been drawing on Lodge's work in his own experiments. When Lodge performed his wireless experiments at the Royal Institution in 1889, the 'walls of the lecture-theatre, which were metallically coated, flashed and sparkled, in sympathy with the waves which were being emitted by the oscillations on the lecture-table'. When he repeated the show for the Society of Telegraph Engineers, the 'same sparkling of the walls also occurred', and a white-faced caretaker came to tell him that 'the gas- and water- pipes in the basement were sparking into each other – a phenomenon that Lord Kelvin and others went down to see'.[46]

A few years later, William Crookes was clearly fully alert to the possibilities of wireless electricity. There could be no more doubt about the ether: 'The researches of Lodge in England and of Hertz in Germany give us an almost infinite range of ethereal vibrations or electric rays, from wave-lengths of thousands of miles down to a few feet.' He was alert to the commercial prospects too. Their experiments 'revealed the bewildering possibility of telegraphy without wires, posts, cables, or any of our present costly appliances'.[47] In fact, Crookes said, he had already seen wireless telegraphy in action. He was referring to experiments carried out more than a decade previously by David Hughes, inventor of the printing telegraph and the microphone. Attempting to improve the microphone's performance, Hughes had been continuing to experiment at his house on Great Portland Street in London. During his experiments, he noticed that they caused a faulty telephone to produce sparks. After further research, he could produce a spark from further down the street. He showed the phenomenon to friends such as William Robert Grove and William Henry Preece (two fellow Welshmen) – as well as Crookes – but was discouraged by George Gabriel Stokes who maintained that nothing novel was going on, and that it was simply an effect of induction.

It was Preece that Marconi went to see as soon as he arrived in London with his wireless telegraph. He had already applied for a patent, but he knew that he needed patronage as well. Preece was soon convinced that Marconi's apparatus had the potential to be the basis of a working system. Within a month he was transmitting from the roof of the Post Office building in central London to another building several hundred yards away. By the end of the year, Marconi was demonstrating his wireless telegraph to the War Office on Salisbury Plain. Preece was certain by now that the young man was 'a great and valuable acquisition'.[48] Marconi's experiments continued with his support. A few years later in May 1897, Marconi succeeded in transmitting wireless signals from the island of Flat Holm in the middle of the Bristol Channel to Lavernock Point on the Welsh mainland. It was touch and go to begin with, and the attempt almost didn't come off: 'On the 11th and 12th his experiments were unsatisfactory – worse, they were failures – and the fate of the new system trembled in the balance. An inspiration saved it. On the 13th the receiving apparatus was carried down to the beach at the foot of the cliff, and connected by another 20 yards of wire to the pole above, thus making a height of 50 yards in all. Result, magic! The instruments, which for two days failed to record anything intelligible, now rang out the signals clear and unmistakable, and all by the addition of a few yards of wire! Thus often, as Carlyle says, do mighty events turn on a straw.'[49]

A few days later, Marconi successfully transmitted signals between Lavernock and Bream in Somerset on the other side of the Bristol Channel. 'Marconi messages have been sent between Penarth and Bream Down, near Weston-Super-Mare, across the Bristol Channel, a distance of nearly nine miles,' reported the *Daily News*.[50] The experiments proved decisive, and it was not long before the press were reporting on Marconi's exploits in glowing terms. It was clear, they reported, 'that the investigations of the last few years have opened up a region which will yield far more remarkable phenomena than any now within the knowledge of man'.[51] The

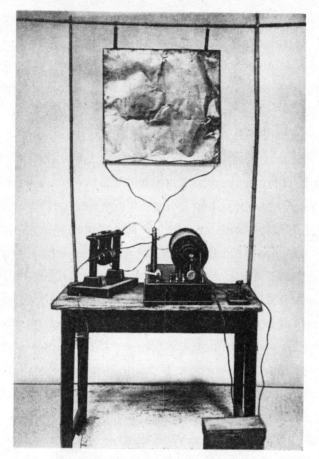

Marconi's first wireless telegraph apparatus.
Radio Broadcast magazine, 1926, 10

success at transmitting messages over expanses of water was the crucial element. This was what had captured the Admiralty's attention. There was nothing new, after all, about transmitting messages overland – the telegraph and the telephone performed that task perfectly adequately. There was nothing new about transmitting messages between different parts of the far-flung Empire, either – that was what the all-red line was about. What was still missing was the capacity to transmit effectively to ships at sea. That was what the experiments at Lavernock were designed to test. If Marconi could send messages between land and an island, then he could send messages to ships as well. It was another little link in the chain of command that governed the Empire.

Others were soon experimenting as well. Adolf Slaby, who had assisted Marconi at Flat Holm, was soon carrying out his own experiments back in Germany. On the other side of the Atlantic, Nikola Tesla had his own ambitious plans for wireless telegraphy. The potential of his experiments to deliver a system of wireless telegraphy to rival Marconi's had formed a key element in his efforts to persuade J.P. Morgan to finance his experiments at Wardenclyffe. In fact, it was clear that the potential to deliver a rival wireless telegraph that did not breach Marconi's patents was the main reason for Morgan's interest in Tesla's experiments. Tesla himself was dismissive of Marconi's efforts. 'While Signor Marconi's experiments are most interesting, they are not novel,' he sneered. 'I do not wish to detract from Signor Marconi's success, but I do not see why there need be much stir about it.'[52] Tesla boasted that with his wireless system he would be able to not only send messages through the ether, but control battleships at a distance. With this revolutionary new weapon, 'war would be abolished', and it would 'work a revolution in the politics of the whole world'.[53] Like Szczepanik, Tesla promised his invention would be on show at the Paris *Exposition Universelle* in 1900. In the meantime, in March 1899, Marconi succeeded in sending wireless messages between England and France.

Tesla, however, had aspirations to send his wireless signals far further than that. He wanted to talk to Mars. The astronomer Percival Lowell's observations of the Martian canals, and the success of H.G. Wells' *The War of the Worlds* in 1898 meant that Mars was very much in the mind of the public during the second half of the 1890s. In 1896, Tesla told newspapers that he had a 'scheme under consideration for five or six years' to communicate with the red planet: 'I am becoming more convinced every day that it is based upon scientific principles and is thoroughly practicable.'[54] A few years later, he repeated that he had 'apparatus that can accomplish it beyond any question'. Whenever he was ready to do so, he could be 'perfectly certain that the electrical effects would be thrown exactly where I desire to have them and that the exact signals I desire to make would be made.'[55] It was not long before he was hinting that he had already received Martian messages. 'I have observed electrical actions which have appeared inexplicable,' he claimed, 'faint and uncertain though they were, and they have given me a deep conviction and foreknowledge that ere long all human beings on this globe, as one, will turn their eyes on the firmament above with feelings of love and reverence, thrilled by glad news.'[56]

With Tesla beating the drum for extraterrestrial communication, writers of scientific romance could allow their imaginations to run riot. In 1903, the American geologist and curator at the American Museum of Natural History, Louis Pope Gratacap, published *The Certainty of a Future Life on Mars*. The novel's conceit was simple. It featured a duo of inventors, father and son, experimenting with Marconi-like apparatus. As well as ardent experimenters, they were devotees of spiritualism. They speculated as they experimented with the notion that the various planets of the Solar System represented different stages in spiritual development. They talked about 'the union of our world with others by magnetic waves, but as it slowly assumed a theoretical certainty he talked more and more boldly of this portentous and transforming possibility'. After the father's death, the son began to receive mysterious signals on his wireless apparatus. The signals transpired to be from his father, confirming that their speculations were true and giving an account of his afterlife on Mars. The novel's theosophical speculations were interspersed with references to the latest technologies. Their laboratory was 'lit by electric lamps constructed somewhat on the principle of Edison's, but using platinum wires'; they discussed the latest astronomical discoveries; and the 'magnetic waves received at first by us presumably from Mars, and later, as the communications indisputably show, from that planet, were taken upon a Marconi receiver, or what was practically that'.[57]

More prosaically, perhaps, by the time Gratacap published his fantasy, Marconi had already succeeded in transmitting wireless messages across the Atlantic. On 12 December 1901, Marconi was successful in sending a wireless signal from his experimental station in Poldhu, Cornwall, to Newfoundland. A few days later, *The Times* made it public: 'Signor Marconi authorizes me to announce that he received on Wednesday and Thursday electrical signals at his experimental station here from the station at Poldhu, Cornwall, thus solving the problem of telegraphing across the Atlantic without wires.'[58] Interviewed in *The New York Times*, H. Cuthbert Hall, Marconi's New York representative, made it quite clear that they meant business. They were now in commercial competition with the telegraph companies. 'Plans were formulated some time ago in anticipation of the successful outcome of Signor Marconi's experiments,' he told them, 'but I do not care to make them public just now.' It took rather longer than Marconi had anticipated, however. Transatlantic wireless telegraphy remained a chancy and delicate affair for several years. By 1904, nevertheless, Marconi was in business using wireless telegraphy to communicate between ships and shore on both sides of the Atlantic. It was not until 1907 that a reliable wireless telegraph service across the Atlantic was established.

In a striking and revealing parallel with the telegraph 65 years earlier, capturing a fugitive criminal played an important role in popularising wireless telegraphy too. In 1845, the part played by the telegraph in bringing John Tawell to justice following his murder of Sarah

Hart was a key event in cementing its reputation as a groundbreaking technology. Similarly, in 1910, the part played by wireless telegraphy in apprehending Dr Hawley Harvey Crippen was a key event in making its reputation too. Following his wife's disappearance from their house at the end of January 1910, a panicked Crippen fled with his lover to Antwerp, where they boarded the SS *Montrose*, bound for Canada. Despite being disguised, they were recognised by the captain. Since the ship was equipped with one of Marconi's wireless telegraphs, just before it went out of range of the transmitter, he was able to send a message to the police: 'Have strong suspicions that Crippen London cellar murderer and accomplice are among saloon passengers.' A detective boarded the faster ship SS *Laurentic* at Liverpool, arrived in Canada before the fugitives and arrested Crippen as he disembarked. That he was the first murderer captured by wireless made Crippen and his trial even more of a sensation. The episode hammered home the power of wireless. This was a technology that really could go anywhere.

From the telegraph to wireless telegraphy, Victorians recognised that new communication technologies were transforming their world, and transforming them in the process. The invocation of magic was telling. Victorians often adopted the language of wonder and enchantment to talk about technology. It was a way of coming to terms with their new power. It was a way of making it familiar. So, the telegraph was 'a spirit like Ariel'. Witnessing wireless telegraphy, one journalist gushed that the 'magic of it all naturally impresses one who for the first time sees the little there is to see.'[59] The connection drawn between wires and nerves was just as telling. The telephone brought people together in just the same way that 'the various members of the body now are by the nervous system'. The metaphor – just like stories about the capture of escaping fugitives – was a recognition of what these technologies offered in terms of surveillance and control. The telegraph and its descendants fed the

Victorians' fantasising about the future. They were glimpses of the world to come. In many ways, the Victorians' culture of technical expertise was intimately bound up with these communication technologies – the telegraph, the telephone and wireless could not exist without armies of technicians. Ironically, they helped nurture the myth of the lone inventor too.

But the invocation of magic underlines the imperialism that was built into the telegraph and its successors. The telegraph, said Edward Copleston, Bishop of Llandaff, 'far exceeds even the pretended magic and the wildest fictions of the East'.[60] It was an arrogant expression of cultural domination. Men like Copleston were supremely confident that they could achieve what others could only pretend. The telegraph was Victorian exceptionalism made tangible with copper and gutta-percha. When the Marquess of Salisbury explained to the revellers at the Institution of Electrical Engineers' annual dinner that they were members of a profession made for empire, he was telling them something they already knew very well. It was on their laboratories and workshops that 'the huge belligerent power of modern States, which marks off our epoch of history from all that has gone before, must be held, by any who investigates into the causes of things, absolutely to depend'.[61] The kind of expert knowledge they and others like them represented was inextricably intertwined with the business of empire. England's empire, just like those of other European powers, and the growing United States on the other side of the Atlantic, depended increasingly on the disciplined and regimented expertise that built and maintained its networks of communication. They were products of applied accuracy and precision. Building and maintaining those networks required the resources that only imperial reach could provide. Telegraphs – and the physics built around them – needed empires as much as modern empires needed telegraphs.

Chapter 7

Calculating People

O ne of the defining memories of Charles Babbage's childhood was being taken by his mother to visit 'exhibitions of machinery'. One in particular stuck in his mind 'by a man who called himself Merlin'. His favourite exhibits at this particular show were 'two uncovered female figures of silver, about twelve inches high'. One of them 'walked or rather glided along a space of about four feet, when she turned round and went back to her original place'. The gracious automaton 'used an eye-glass occasionally, and bowed frequently, as if recognising her acquaintances'. As for the other silver figure, she was 'an admirable danseuse, with a bird on the fore finger of her right hand, which wagged its tail, flapped its wings, and opened its beak'. The silver dancer 'attitudinised in a most fascinating manner', and her 'eyes were full of imagination, and irresistible'.[1] Merlin was, in fact, John Joseph Merlin, an instrument maker born in Wallonia in Belgium who had arrived in London in the Spanish ambassador's train in 1760. He was famous for his automatic mechanical wonders, including a self-playing Welsh harp. Babbage was probably about ten or so when he visited Merlin's Mechanical Museum to be entranced by the two beguiling automata. In later life, he came across the second of the two seductive silver ladies in an auction and promptly bought her.[2]

The two charming mechanical performers stuck in Babbage's mind, perhaps, because they were such good examples of machines mimicking humans. He would not have been the first to be struck in this way. Two centuries earlier, the philosopher René Descartes had turned to automata as a way of defining what it meant to be human. Animal and human bodies were just like automata, according to him. What made humans different from animals was that they were automata that contained a soul, while animals really were just machines. Automata were fashionable. The French mechanic Jacques de Vaucanson, for example, made himself famous as a maker of astonishingly intricate and complex automata for his wealthy patrons. Among his most famous creations were a flute player, a tambourine player and, astonishingly, an automaton duck. The duck contained more than 400 moving parts and could even apparently eat, digest and defecate.[3] Seeing devices like these, and reading Descartes with a subversive eye, led the radical French philosopher Julien Offray de La Mettrie to the conclusion that Descartes had not gone far enough: humans were not automata with souls, they were simply automata. There was no such thing as a soul, and the mind was nothing more than the product of complex machinery – a mere secretion of the brain.

By the end of the eighteenth century, ingenious mechanics did indeed seem capable of constructing clockwork minds, as well. Wolfgang von Kempelen toured Europe from 1770 onwards, showing off his chess-playing Turk. Von Kempelen spent most of his career as a court official for the Hapsburg Empire, but he was also an enthusiastic contriver of ingenious machinery. He had designed a manually operated speaking machine, for example. The Turk had been built originally as a gift for the Empress Maria Theresa. The automaton consisted of the figure of a Turk sitting in front of a chess set. It responded to moves from an opponent by moving its arm. The cabinet beneath the chess set was opened as part of the performance to demonstrate to the audience that there was no room for anyone to hide inside the machine (though, as it turned out, there was, and the Turk was in fact operated by a hidden player). To all appearances, the mechanical Turk was playing chess on its own. For more than eight years, first von Kempelen, then a string of other showmen, after his death, took the Turk on tour, challenging their audiences to play against it – and discover how it worked.

Wolfgang von Kempelen's chess-playing Turk.
Carl Friedrich Hindenburg, *Ueber den Schachspieler des Herrn von Kempelen, nebst einer Abbildung und Beschreibung seiner Sprachmachine*
(Leipzig: J. G. Müller, 1784)

Babbage was certainly familiar with the automaton – and was himself an enthusiastic chess player.

Preparing himself for student life at Cambridge, Babbage recalled in his autobiography that he had 'formed a plan for the institution among my future friends of a chess club, and also of another club for the discussion of mathematical subjects'.[4] Chess was a passion throughout his undergraduate years. Games with fellow aficionados were trials of intense concentration. A 'fellow-commoner at Trinity named Brande' was a favourite adversary. They would sit in complete silence as they played and discussing the match afterwards would reconstruct from memory the state of the board at different stages in their game. Memory, of course, was the key. Babbage found in later life that when he played against this particular player, his only hope of victory 'was by making early in the game a move so bad that it had not been mentioned in any treatise', since Brande possessed and had memorised them all.[5] 'Brande' was, in fact, John Brand, who was a prominent London chess (and cricket) player throughout the 1820s. He was one of a coterie of dedicated players who to a large extent built their lives around their passion for the game. He was a stiff opponent for Babbage – and Babbage must have been no mean player himself to have had any chance of challenging him.

Brand was a member of the London Chess Club, and in 1820, when Johann Maelzel, then the owner of von Kempelen's chess-playing Turk, brought the automaton to London, Brand was one of the players selected to challenge him. He even succeeded in beating the Turk a few times.[6] His capacity to hold his own against the automaton consolidated his reputation as one of London's leading exponents of the game. He took part in a famous tournament between Edinburgh and London players in 1824. In 1830, Brand was declared insane and spent the rest of his life in an asylum. One of the symptoms of his madness was that he was convinced that he had paid £1,500 for a mechanical chess player. The game had clearly become an obsession for Brand, but before his mind broke down, he exemplified the appeal chess had for Babbage. There was something mechanical about the game – and hence the attraction. It required a very specific kind of intelligence. The newly established medical journal *The Lancet* included a chess column, recommending it to its readers on the grounds that 'it strengthens the intellectual faculties, by the unremitting attention which it demands' and 'the lessons of foresight, patience, and perseverance which it inculcates'.[7] Chess made thought automatic and methodical.

Mathematics was just the same. By the time Babbage arrived at Trinity College, Cambridge, in October 1810, he was already 'passionately fond of algebra'. He was familiar already with the latest French analytical approaches and was dismayed to find that Cambridge was still mired in old-fashioned Newtonian approaches to the subject. Eventually, in 1812, he and a handful of like-minded enthusiasts established the Analytical Society. It was, at least in part, meant to be a joke. At the time, Cambridge was riven by a debate about whether or not Bibles should be annotated to explain what obscure passages really meant. A Cambridge branch of the British and Foreign Bible Society had recently been established with a mission to make Bibles more widely available. The debate raged between high churchmen who wanted the Bible distributed along with the Anglican Book of Common Prayer and evangelicals who were scandalised at any attempt to improve upon the word of God. Babbage and his friends thought the debate was ridiculous, and the founding of the Analytical Society was meant to be an attempt to poke fun at it. It was their idea of parody.

The Analyticals' plan, accordingly, was for 'a society to be instituted for translating the small work of Lacroix on the Differential and Integral Calculus. It proposed that we should have periodical meetings for the propagation of d's; and consigned to perdition all who supported the heresy of dots. It maintained that the work of Lacroix was so perfect that any comment was unnecessary.'[8] When it came time to publish their proceedings, Babbage playfully suggested 'The Principles of pure D-ism in opposition to the Dot-age of the University' as a title. It was a skit on the biblical argument as much as a jab at the outdated Cantabrigian reverence for Newtonian mathematics. It was deadly serious as well, though. The holy scripture at issue here was *An Elementary Treatise on the Differential and Integral Calculus*, written a few years earlier by the French mathematician Sylvestre François Lacroix, who was one of the leading exponents of the new approach, which adopted Gottfried Leibniz's more versatile notation for the calculus, rather than the old-fashioned and clunky Newtonian fluxions still in patriotic use at Cambridge. Babbage and his fellow Analyticals were convinced that the new French methods offered a path to mechanising mathematics – that highest of intellectual pursuits.

But Babbage wanted more. 'I wish to God these calculations had been executed by steam,' he fumed at John Herschel, in the summer of 1821.[9] The two friends had been checking tables of calculations carried out by computers (young men hired to carry out tedious and repetitive arithmetical work) to compile astronomical tables for the Astronomical Society. Babbage was expressing his frustration at the number of errors they found. It was the germ of an idea for the young man obsessed with automata, chess and French mathematics: maybe calculations really could be carried out by steam-driven machinery. Computers – that is to say, people who were paid to carry out tedious and repetitive calculations – did important work. The young men who laboured in actuaries' offices and astronomical observatories to calculate the tables on which businessmen, astronomers and ships' captains depended were essential cogs in the machines. Many of the Astronomical Society's fellows combined financial work with astronomical interests. As far as they were concerned, it was natural that science and commerce should operate along the same lines – that was why so many of them were deeply involved in the campaign to reform the Royal Society. Their reforms, as they saw them, were aimed at making the business of science more efficient – and more profitable. To men like these, and to Babbage, it seemed clear that replacing less than reliable human intelligence with the more efficient (and uncomplaining) intelligence of machinery was simply good business.

The potential use of thinking machines as replacements for thinking humans was part of Babbage's plans from the very beginning of his speculations. It was exactly what steam calculation was designed for. As he explained to Humphry Davy, embattled president of the Royal Society, it was the 'intolerable labour and fatiguing monotony of a continued repetition of similar arithmetical calculations, first excited the desire, and afterwards suggested the idea, of a machine, which, by the aid of gravity or any other moving power, should become a substitute for one of the lowest operations of human intellect'.[10] Using as his model the monumental set of tables calculated in France under the supervision of the eminent mathematician Gaspard de Prony, Babbage showed Davy that deploying a calculating engine such as the one that he envisaged would reduce the number of human calculators required from 96 to twelve.[11] Prony's treatment of the whole process of calculating the tables as if the human calculators were elements of a machine was vital for Babbage's thinking on this, too. Prony argued that what mattered for efficient computing was not the intelligence of the individual computers, but the intelligence of the system as a whole. That was just what Babbage wanted to achieve with his engine. It was meant to replace a whole group of skilled workers, not just individuals.[12]

This was a hugely ambitious project – and it would not be cheap. That was why Babbage wrote to Davy in the first place, after all. The letter was an opening salvo in his campaign to persuade the Royal Society to lobby government on his behalf for the substantial sums of money he would need to build what he called the Difference Engine. His efforts to fund the engine were taking place, of course, against the background of the increasingly virulent campaign for Royal Society reform and his own prominent role in that business. This was not a coincidence, either. One of the reasons for Babbage's increasing dissatisfaction with the state of the Royal Society was what he saw as its reluctance to properly support the grand design for mechanising the work of the mind. Despite the fact that the Royal Society did persuade a sceptical government to pour money into the project (starting with an initial investment of £1,700), Babbage was convinced that they could – and should – do more. The Difference Engine would be the future of efficient calculation in the national interest, doing away with the uncertainties of merely human mental labour. It would be a far better use of the Royal Society's resources, Babbage thought, than doling out comfortable sinecures to cronies.

It was a tremendous and challenging undertaking. The very idea of a thinking machine seemed fantastical. When John Millington, the Royal Institution's professor of mechanics, delivered a lecture on Babbage's vision in 1823, a sceptical correspondent from the *Morning Chronicle* suggested that 'were it not for the credit we feel inclined to give to the veracity of Professor Millington, who stated that he had seen the machine at work, we should almost feel inclined to doubt the existence of an engine possessing such extraordinary powers as he ascribed to that of Mr Babbage'.[13] Millington, of course, had not actually seen the whole machine in action, since, other than in Babbage's brain, only a small model actually existed at that time. Getting it into the real world was going to take more than words.

As Babbage's friend John Herschel put it, defending him from his detractors, 'such is its extent, such the variety of mechanical movements to be contrived and executed, and such the elaborate perfection of workmanship which has been found necessary to bestow on all its parts, to afford a moral security for its successful action when

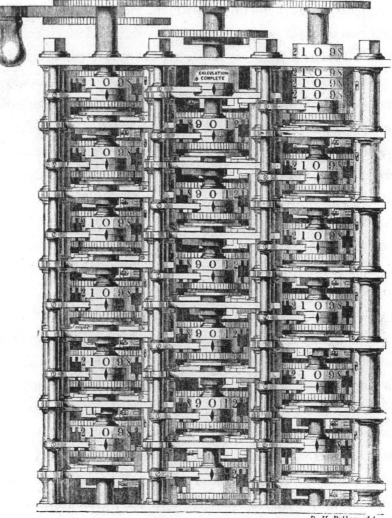

B. H. Babbage, del.

Part of the Difference Engine.
Charles Babbage, *Passages from the Life of a Philosopher*
(London: Longman & Co., 1864)

put together, that a very long time must yet elapse, and a very heavy further expense be incurred, before it can be completed'.[14]

That, though, was just where dreams about mechanical minds clashed with mechanical reality – and the politics of mechanics. Babbage wanted the best, and this meant the best mechanic, working on his engine. That mechanic was Joseph Clement, who was widely recognised as one of the finest toolmakers in the country. Babbage hired him to build the engine and to design and make the precision tools needed to do the job. This was where cultures clashed. Clement was well aware of his value – and charged accordingly. He was also well aware of his rights. According to the custom of his trade, tools belonged to the craftsman who used them. As far as he was concerned, the tools that he had designed and made to build the Difference Engine belonged to him. Babbage took the view that since he had paid for them, they were his property. But Babbage also knew that Clement was indispensable – 'to preserve the life of Mr Clement' was the 'first necessity' of building the engine – 'it would be extremely difficult if not impossible to find any other person of equal talent both as a draftsman and as a mechanician', he admitted.[15] When Clement walked away from the project in 1833, taking his tools with him, it spelled the end for the Difference Engine.

But Babbage was disenchanted with his plan by then in any case. His fantasies of artificial minds had grown far more ambitious since he first started dreaming about steam calculation a decade earlier. The Difference Engine had been a potent symbol of the ways in which the disciplined and directed vision of science he was even then trying to impose on the Royal Society could transform the future. It opened up a vista of automatic machinery toiling ceaselessly – mechanical minds and mechanical bodies working together. When Andrew Ure, author of *The Philosophy of Manufactures* rhapsodised about the ideal factory as 'a vast automaton, composed of various mechanical and intellectual organs, acting in uninterrupted concert for the production of a common object, all of them being subordinated to a self-regulated moving force', he might almost have been thinking of Babbage's machine.[16] It shows, at any rate, how very much of its time the vision was. Babbage by now was intent on exploring the possibilities of an artificial mind even further, though. The Difference Engine had been designed to carry out a very specific and limited set of intellectual tasks, mechanising the kinds of routine calculations that had frustrated Babbage so much as a younger man. Now, however, he was convinced that machine minds were capable of so much more.

Memory and foresight

Even without being built, Babbage's engine was a sensation. Dionysius Lardner was hugely impressed by the ambition 'to reduce arithmetic to the domination of mechanism, to substitute an automaton for a compositor, to throw the power of thought into wheelwork'.[17] It was astonishing that such a thing could be 'the production of solitary and individual thought begun, advanced through each successive stage of improvement, and brought to perfection by one mind'.[18] Lardner had reasons of his own for this enthusiasm for Babbage's project to build a mechanical mind. He, too, was fascinated by the prospect of understanding the mind as if it were a piece of machinery. This was why he was so taken by the latest philosophical craze of mesmerism. Just as he was writing about Babbage's engine, he was also speculating excitedly about the prospect that mesmerism (what would now be called hypnotism) – and the idea that body and mind might communicate with each other through a force something like magnetism – might lead to a new understanding of the way the human mind worked. He was sure that the mechanical notation that Babbage had developed to describe the Difference Engine could be applied to describing all the functions of human and animal bodies as well – 'all would find appropriate symbols and representatives in the notation'. It was as if the mechanical language Babbage had designed for his engine 'had been designed expressly for the purposes of anatomy and physiology'.[19]

It was also, Lardner noted, well designed to describe the operations of the modern factory. The readers of this paean of praise for Babbage's engine in the highbrow *Edinburgh Review* could not have failed to have been struck by the way in which Lardner's meticulous description of the way the device worked by breaking calculation down into its essential parts was redolent of the factory system. In case they failed to see it, Lardner made the point clear in any case. The mechanical notation might be used to 'exhibit, in the form of a connected plan or map, the organisation of an extensive factory, or any great public institution, in which a large number of individuals are employed'. It might even suggest ways of improving efficiency.[20] Babbage knew perfectly well that his engine worked like a factory. That was what it was for. When he published *On the Economy of Machinery and Manufactures*, he was clear that it was 'one of the consequences that have resulted from the Calculating Engine, the construction of which I have been so long superintending'. To refine his plans, over the past decade Babbage had visited 'a considerable number of workshops and factories, both in England and on the Continent, for the purpose of endeavouring to make myself acquainted with the various resources of mechanical art', and he had been 'insensibly led to apply to them those principles of generalisation to which my other pursuits had naturally given rise'.[21]

His book was a meticulous and thoroughgoing survey of the factory system, highlighting the ways in which the division of labour – reducing complex tasks to their simplest components – made production more efficient. This was the principle underlying the production of mathematical tables, too, and Babbage's engine was meant to work by embodying it in machinery. Taken to its logical conclusion, the division of labour meant that 'the master manufacturer, by dividing the work to be executed into different processes, each requiring different degrees of skill or of force, can purchase exactly that precise quantity of both which is necessary for each process; whereas, if the whole work were executed by one workman, that person must possess sufficient skill to perform the most difficult, and sufficient strength to execute the most laborious, of the operations into which the art is divided'.[22] Factories made money – and Babbage's engine made numbers – by breaking things down to their essential components. Just as factories needed workers to carry out the same simple task over and over again, the various components of the Difference Engine were designed to carry out just one part of a complex calculation over and over again. This was intelligence stripped down to its basics and built into a machine.

By the 1830s, as Babbage battled with Joseph Clement, pondered about the economics of factories and the division of labour and recovered from the vicious battle for control of the Royal Society, he came to realise that the intelligence embodied in the Difference Engine's cogs and gears was a very limited one, after all. Disenchanted, he started to dream about the possibilities of a far grander design – something that he would eventually call the Analytical Engine. Where the Difference Engine could only perform a very limited number of mathematical tasks, the new device would be far more versatile. It would embody an intelligence far more human. It all boiled down to Babbage's understanding of what the bare bones of intelligence really were. Intelligence was made of memory and anticipation. That's what was needed for mastery in a game of chess, for example. The winning player was the one who (like his old friend Brand) could remember all past moves and combinations and could anticipate the future progress of the game. A machine that could do that would, unlike von Kempelen's chess-playing automaton (which actually concealed a cunningly hidden human player), really be intelligent. This was the Analytical Engine. Memory would be embodied in the machinery that registered all past calculations. Anticipation was the problem.

The key to Babbage's conundrum was another piece of precision industrial engineering. The Jacquard loom had been invented by the French weaver Joseph Marie Jacquard and patented in 1804. It was a device that made it easier to weave complex patterns on a loom, using a series of punched cards to control the loom's mechanism. Something similar had been proposed a few decades earlier by Jacques de Vaucanson, who was more famous for his construction of elaborate automata. Jacquard's loom revolutionised the French weaving industry – and set Babbage thinking about how he could instruct his projected Analytical Engine. Babbage would certainly have encountered Jacquard looms during his factory tours, but even if he had not he would have seen the one on show at the Adelaide Gallery. As Ada Lovelace put it, 'the Analytical Engine *weaves algebraical patterns* just as the Jacquard-loom weaves flowers and leaves', and 'those who may desire to study the principles of the Jacquard-loom in the most effectual manner, viz. that of practical observation, have only to step into the Adelaide Gallery or the Polytechnic Institution. In each of these valuable repositories of scientific *illustration*, a weaver is constantly working at a Jacquard-loom, and is ready to give any information that may be desired as to the construction and modes of acting of his apparatus'.[23]

The Analytical Engine did not just model the human mind, though. It offered insights into the mind of God as well – and offered a solution to the problem of miracles. In his *Ninth Bridgewater Treatise*, Babbage suggested that the Analytical Engine's operation showed how miracles – events that appeared to contravene the laws of nature – could occur without actually contravening natural law. Suppose, he argued, that someone saw the engine carrying out its calculations, adding two to a figure each time. They would eventually conclude that this was the law according to which the engine operated, and when the series of figures suddenly changed, they would be unable to explain it. But what had happened was that the engine had been arranged to generate figures in one way up to a certain number of times, but then to proceed differently. That was what miracles were, Babbage suggested. They were not really violations of natural law, but expressions of a deeper and unrecognised one.[24] The example showed how important Babbage thought the engine and the principle it embodied was. As far as he was concerned, it really did represent the distilled essence of intelligence. It revealed just how far the mechanisation of mind might go – and just how powerful those mechanical minds could be.[25]

There was a strong streak of radical politics in all this as well, of course. If machines could think, then minds were matter, and if there were souls then they were mechanical, too. Thomas Simmons Mackintosh, a follower of the utopian socialist Robert Owen, argued, for example, that all this was evidence of the futility of Christianity. 'It is better to view man as an organised machine, and to search for the seat of those impulses in the functions of his physical nature, where assuredly they are to be found,' he argued, 'than to trace them to sources beyond our knowledge and above our control.'[26] Mackintosh thought that electricity was the fuel that ran the human machine – and this was a common view among radicals. Their confidence in the body electric was bolstered by experiments that seemed to show how electricity could (almost) restore the dead to life. In 1803, Giovanni Aldini had experimented on the corpse of an executed murderer. Andrew Ure did the same in 1818.[27] One account of Ure's experiments described how 'the corpse seemed to point to the different spectators, some of whom thought it had come to life!'[28] Radical as he was, Lord Byron could not stop himself from poking fun at some of these ideas, quipping in *Don Juan* that 'galvanism has set some corpses grinning'.[29]

Byron's daughter, Ada Lovelace was one of Babbage's key allies in getting his plans for the Analytical Engine into the public eye. Ada, the Countess of Lovelace, Lord Byron's only legitimate daughter, although he never met her, was brought up by a mother determined to make sure that she did not follow in her father's dissolute footsteps. Annabella Byron chose mathematics as the instrument for disciplining her daughter's impressionable young mind. She hired mathematical tutors to provide the 'watchful and judicious superintendence' she thought her daughter needed to 'form the basis of good habits'.[30] She clearly agreed with Babbage that mathematics was the sure foundation of a disciplined mind. Ada agreed that 'nothing but very close and intense application to subjects of a scientific nature now seem at all to keep my imagination from running wild',[31] and her tutor agreed, telling her that mathematics was a 'moral discipline, tending to control the imagination, and give one mental self-command'.[32] Lovelace was a familiar figure in fashionable London by the beginning of the 1830s. Her aristocratic and intellectual upbringing (as well as her relationship to the notorious Byron) gave her an easy entry to London's scientific circles. She was soon rubbing shoulders with people like Mary Somerville and the London University's mathematical professor, Augustus De Morgan. Sooner or later, she was bound to meet Babbage.

In fact, she first met him at a soirée on 5 June 1833, when she was only seventeen. A few weeks later, her mother recorded how they 'both went to see the thinking machine (for so it seems) last Monday'. Annabella confessed that she 'had but faint glimpses of the principles by which it worked', but could still recognise that 'there was a

sublimity in the views thus opened of the ultimate results of intellectual power.'[33] Ada was fascinated by its possibilities and was soon
a frequent guest at Babbage's own fashionable soirées. She was an
enthusiastic recipient of his speculations as he plotted the Analytical
Engine during the 1830s and 1940s. As Byron's daughter, she was
well placed in London society circles to help spread the message.
Babbage toured Italy in 1840, giving lectures to men of science and
government officials on the potential of his engine. One convert to
his schemes was Luigi Federico Menabrea, engineer and later prime
minister of Italy. Menabrea was enthusiastic enough to write a long
panegyric on the engine, and it was to Lovelace that Babbage turned
to have it translated into English. 'The engine,' Menabrea declared,
'may be considered as a real manufactory of figures.'[34] Lovelace
added her own extensive notes to the translation, offering detailed
examples of how the device would operate. She even claimed that
the Analytical Engine might be able to compose music, so convinced
was she of its possibilities.

The Analytical Engine remained even more of a dream than the
Difference Engine – at least a portion of that had been built by
the 1830s, though it was never completed. The machine retained its
grip on the popular imagination for the rest of the century though.
It stood for what mechanical intelligence might be. As late as 1878, a
committee of the British Association for the Advancement of Science
deliberated over whether the time had come to make the Analytical
Engine a reality. It was 'a marvel of mechanical ingenuity and
resource', the committee concluded, acknowledging that 'the existence of such an instrument would place within reach much which,
if not actually impossible, has been too close to the limits of human
skill and endurance to be practically available.'[35] Even 40 years after
its inception in Babbage's mind, though, they were not sure that the
technology existed to actually build it. It remained a potent symbol of

what might be. Harry Buxton, Babbage's biographer, marvelled that, in the Analytical Engine, the 'marvellous pulp and fibre of a brain has been substituted by brass and iron, he has taught wheelwork to *think*, or at least to do the office of thought'.[36] Following his death in 1871, Babbage's own brain was preserved and was eventually examined on behalf of the Royal Society in 1907.[37]

In 1879, just a year after the BAAS committee's report (and maybe inspired by it), the *New York Sun* featured a short story by Edward Page Mitchell, one of the newspaper's journalists (and later its editor), called 'The Ablest Man in the World'. The *Sun* published stories like this, cleverly positioned on the boundary between fact and fiction, quite regularly. The story featured the Baron Savitch, the greatest diplomat in Europe, and his doctor, Rapperschwyll. The story hinged around the American narrator's discovery that Rapperschwyll had started life as a clockmaker, fascinated by Babbage's Analytical Engine. He had succeeded in constructing 'a machine that went far beyond Babbage's in its powers of calculation', and when fed with facts, produced conclusions. The machine 'eliminated the personal equation; it proceeded from cause to effect, from premise to conclusion, with steady precision'. Rapperschwyll had become obsessed with the prospect of taking an individual, 'removing the brain that enshrines all the errors and failures of his ancestors back to the origin of the race' and replacing it 'with an artificial intellect that operates with the certainty of universal laws'. Baron Savitch was the result of his obsession – an imbecile whose brain had been replaced by this perfect mechanism.[38] It was a play on the possibilities that the Analytical Engine were still seen to offer.

Calculating empires

By the end of the century, state bureaucracies were certainly faced with an avalanche of numbers and were in desperate need of a technology for dealing with them. They would have had little difficulty in finding plenty of uses for a diplomat with a mind like Savitch's, so Mitchell's choice of occupation for his calculating machine-man was a telling one, and one that would have resonated with his readers. Late Victorian diplomats, of all people, were men who needed to calculate finely. Even as Babbage designed his engines, the tables they were meant to generate were as likely to be actuarial as astronomical. Babbage himself was very much an enthusiast for statistics. He was a key figure in establishing the statistical section of the British Association for the Advancement of Science. Thomas Malthus, author of *An Essay on the Principle of Population*, arguing that population growth would always outstrip resources (and which was a key resource in Charles Darwin's development of his ideas about natural selection), was another important figure in this respect. He and Richard Jones, his successor as professor of political economy at the East India Company College, were founders of the Statistical Society of London in 1834 along with Babbage, who remained an active member of the society for the rest of his life. It was an article of faith that reducing people to numbers was the secret of efficient government. As the Empire grew, so did the volume of numbers it generated. Its rulers understood perfectly well that ruling meant calculating, and calculating as efficiently as possible. As the city at the heart of imperial power, London had to be a centre of calculation, too.

Statistics' promoters thought that the 'state of our commerce and manufactures, the results of machinery, the effects of free trade, are mere arithmetical problems, more or less involved, that may be worked out if correct data are obtained'. If the numbers were crunched the right way, then 'solutions thus educed should be as certain and as little open to cavil as a proposition in Euclid, or the determination of an algebraic equation'.[39] Others, though, like Cambridge's professor of geology, Adam Sedgwick, were less sure, worried that statistics mixed science and politics too closely. He warned the BAAS leaders that 'if they went into provinces not belonging to them and opened a door of communication with the dreary world of politics, that instant would the foul demon of discord find his way into their Eden of philosophy'.[40] People like Babbage had little truck with that. Statistics was the universal science and its ramifications everywhere: 'Every subject relating to mankind itself, forms a part of Statistics; such as, population; physiology; religion; instruction; literature; wealth in all its forms, raw material, production, agriculture, manufactures; commerce; finance; government; and, to sum up all, whatever relates to the physical, economical, moral, or intellectual condition of mankind.'[41]

Whatever Sedgwick's pious hopes, numbers simply could not be disconnected from their politics. Just deciding what (and who) to count was a political matter that went to the heart of imperial government. From the 1840s onwards, imperial administrators attempted to count the inhabitants of the Empire, their census efforts timed to coincide with the decennial census of the British state itself. Statisticians tabulated populations by age, occupation and ethnicity. Anthropometricians measured skulls and classified people by eye or hair colour. Criminologists collated statistics and information about criminal types, and by the end of the century collected fingerprints. And it was not only the imperial state that needed and generated numbers. Numbers were the lifeblood of commerce as well as state bureaucracies. Banks and insurance companies, manufacturers, railway and telegraph companies all had their tables. Even department stores needed to track the movement of stock. It was this ever-accelerating flow of numbers cascading through the Empire that motivated the BAAS to take a closer look at Babbage's engine and its potential. If dealing with this sea of data could be managed 'by merely selecting or punching a few Jacquard cards and turning a handle, not only much saving of labour would result, but much which is now out of human possibility would be brought within easy reach.'[42]

But while the Analytical Engine might remain a hopeful fantasy, other mechanical calculators really were being built. As early as 1834, a Swedish engineer called Georg Scheutz was already trying to build a simpler version of Babbage's Difference Engine, having first come across it while translating Babbage's *Economy of Machinery and Manufactures* into Swedish. By the end of the 1840s, Scheutz and his son Edvard had succeeded in building a working machine. They came to London in 1855, and thanks to the engineering firm of Donkin & Co., who had made some of the components, their engine was put on show at the Royal Society, with Babbage's own enthusiastic support. It got some distinguished visitors. On 30 June, the *Illustrated London News* reported that 'On Monday morning, about ten o'clock, his Royal Highness Prince Albert, attended by Captain the Honourable D.C.F. De Roos and Dr Becker, inspected the calculating Machine at the apartments of the Royal Society, at Somerset-house, where the Prince was received by Mr Gravatt and Mr Donkin, by whom the machine was explained to his Royal Highness.' It won a gold medal at the *Exposition Universelle* in Paris later that year. A year later, the engine was sold to the Dudley Observatory in the United States. The British government commissioned Donkin & Co. to build another one for £1,200, and when complete it was put to work at the National Register Office, computing the English Life Tables.

There were other – and simpler – devices for calculating and recording numbers on the market. In 1869, Thomas Edison – who would find fame a decade later with his phonograph – patented a device for recording votes electrically, though he failed to find much of a market for it. More successfully, in 1874, a small-town American businessman and inventor, Frank Baldwin, took out a patent for an improved arithmometer that could carry out simple calculations. In 1879, an Ohio saloon owner, James Ritty, patented an automatic cash register, designed to store and display the amount of money paid for each transaction. Later models were patented and marketed as Ritty's Incorruptible Cashier. They were explicitly meant to police the behaviour of shop assistants, since as they recorded the total cash paid in, they made pilfering more difficult. In 1887, another American, Dorr Felt, took out a patent on a device he called the comptometer that used a keyboard similar to that of a typewriter (another late Victorian invention) to put in the numbers. In 1889, he patented the comptograph – a comptometer incorporating a printing mechanism so that results could be printed. Devices like these sold readily on both sides of the Atlantic, finding uses in factories, shops and offices – and even in laboratories.

Different devices had their admirers and detractors. The physicist Charles Vernon Boys thought that comptometers were 'a retrograde step' compared to arithmometers, but 'the knowledge that the comptometer was extensively used in the United States, where appreciation of time-saving appliances is more developed than here ... made me feel that the comptometer must have advantages perhaps more than sufficient to compensate for its operative deficiencies'. He complained that 'the comptometer makes a most aggravating noise, like a typewriter through a megaphone', but recognised that the fact it was operated through a keyboard was a huge advantage in putting in the numbers.[43] Boys might have been thinking about the superior arithmometer invented by the Swedish (though Russian-based) engineer, Willgodt Theophil Odhner, in 1873. His St Petersburg factories had started manufacturing improved versions of the machines on an industrial scale in 1890, and a year later opened another factory in Braunschweig in Germany – marketed across Europe as the Brunsviga (Latin for Braunschweig) calculator.

Advert for a comptometer.
Nature, 1901, 64

When the eugenicist and statistician Karl Pearson, protégé of arch-eugenicist Francis Galton, established his biometrics laboratory at University College, London in 1903, he used Brunsviga calculators. The machines were widely advertised in scientific journals as being ideal for laboratory use.

Inspired, quite possibly, by Babbage's use of punch cards in his Analytical Engine, in 1884, Herman Hollerith took out a patent in the United States for a system that used similar cards to feed information into the machine. Hollerith was a graduate of the Columbia University School of Mines and started experimenting with punch cards while working at the Massachusetts Institute of Technology in 1882. By 1884 he was working for the United States Census Bureau and his machine – the Hollerith Electric Tabulating System – was designed specifically with the needs of census-taking in mind, although as Hollerith pointed out, it could also be used in generating 'the statistics of registration of births, deaths and marriages which are compiled by counting or adding single units as persons in the above'.[44] Ninety-six of Hollerith's machines were bought by the Census Bureau and used during the 1890 US census. As Hollerith explained to the Royal Statistical Society in 1894, 'without the slightest delay such an electrical counting machine will read or test before tabulating whether the given person was white, native born, native father, native mother, male, blacksmith, and resident of New York City'. The machine could even detect errors in the information that was fed into it.[45] They were so astonishingly easy to use, that 'the job can be put in the hands of a girl'.[46]

Machines like these were essential to running bureaucratic affairs by the end of the nineteenth century. That was why Mitchell's picturing of a man with a mechanical mind as a diplomat worked so effectively. He was asking his readers to imagine something like the machinery that ran more and more of the affairs embodied inside a single human head. It was ironic that Mitchell's story ended with its narrator murdering the 'marvel of mechanical ingenuity' that was Baron Savitch since late Victorian life depended increasingly on machinery capable of calculating on an industrial scale. Machines did more than tabulate the census. By the end of the nineteenth century, people could travel by train, send telegraph messages, buy cheap goods or invest for the future because the calculating work that underpinned those things was performed with machines. Thinking, like weaving, was now mechanical. Machines like these raised questions about intelligence, as well, though. Babbage thought that his Analytical Engine would possess the essential elements of intelligence – memory and foresight: just the characteristics needed to play a good game of chess, or solve a complex mathematical equation. Savitch, until one looked inside his skull, at least, was portrayed as not just a highly intelligent man, but a cultured and urbane one as well.

Automatism

Phileas Fogg, the main character of Verne's *Around the World in Eighty Days* was, as his manservant Passepartout noted, a man who did everything like clockwork. He was, Passepartout decided, 'a real machine'. In his fantasy about a living automaton trying – and failing

– to move effortlessly around a world that had been thoroughly disciplined, Verne was drawing on a long history of automata that stretched back to the animated figures of medieval cathedral clocks. His readers would have been familiar with more recent automata such as Jacques de Vaucanson's duck. They would probably have been familiar with von Kempelen's chess-playing Turk and the challenge it posed its audiences to decide whether it was, or was not, a real machine. Automata were part of the culture of scientific exhibitionism, on show at places like the Royal Polytechnic Institution. In 1866, for example, the Christmas season at the Polytechnic was enlivened by the performances of an automaton trapeze artist, based on the real gymnast Jules Léotard (the original 'Daring Young Man on the Flying Trapeze'). Babbage's Analytical Engine raised the prospect of intelligent automata. In 1849, the eccentric electrical experimenter (and surgeon to the Bank of England), Alfred Smee, proposed in his *Elements of Electro-biology* that humans and other animals worked by electricity, and suggested electrical apparatus that mimicked eyes and ears to demonstrate his argument. Sarcastic reviewers suggested that he really should 'construct genuine seeing eyes, capable of being supplied at a low price to the unfortunates who stand in need of them.'[47] Another poked fun at Smee's supposed efforts to build 'a living, moving, feeling, thinking, moral, and religious man.'[48]

This was all part of a culture that wondered (and sometimes worried) about those kinds of possibilities. Would the future really be a world where machines would be difficult to distinguish from people? In his *L'Ève Future*, the French writer Auguste Villiers de l'Isle-Adam played with exactly that conundrum. Predictably, the heroic inventor in the case was Thomas Edison, rising to the challenge of constructing an artificial woman. L'Isle-Adam's Edison – as the author was careful to explain to his readers – was not quite the same Edison as the one who inhabited the real world. Like his real-life counterpart, though, this Edison lived in a world that was thoroughly drenched with electricity. When his old acquaintance Lord Ewald called to see him, he was introduced into a household where everything was electrical. Edison was surrounded by phonographs, telephones and electric lights: all things that looked more and more like tangible emblems of the coming future. When Lord Ewald was first introduced to the android Hadaly, she was described like a 'coat of armour, shaped for a woman out of silver plates'. She was, as the fictional Edison explained, 'nothing at all *from the outside* but a magneto-electric entity'. What Edison proposed was to transform her into a simulacrum of Ewald's unsatisfactory lover Alicia Clary, with all her physical charms, but without her intellectual imperfections.

Jerome K. Jerome, journalist and author of *Three Men in a Boat*, had a more cynical view of automata. He released a story that was a tale within a tale, serialised as 'Novel Notes' in *The Idler*, the magazine he had founded and edited along with Robert Barr a few years earlier in 1892. One of the protagonists of Jerome's story entertained his fellows with the cautionary tale of a village toymaker called Nicholaus Geibel. He could make 'rabbits that would emerge from the heart of a cabbage, flop their ears, smooth their whiskers, and disappear again; cats that would wash their faces and mew so naturally that dogs would mistake them for real cats and fly at them'. Responding to complaints from the village women that none of the men could dance well enough, he made them an automaton dancer. At the next village dance they were introduced to Lieutenant Fritz, who 'keeps perfect time, he never gets tired, he won't kick you or tread on your toes; he will hold you as firmly as you like, and go quickly or slowly as you please; he never gets giddy; and he is full of conversation'.[49] Things did not go well as the automaton dancer performed so efficiently and relentlessly that he danced his partner to death.

Hadaly was not the only female automaton imagined at this time, although maybe E.E. Kellett had her example in mind when he wrote 'The Lady Automaton' for *Pearson's Magazine*. Like Jerome's effort, it was another cautionary tale. The brilliant inventor, this time, was a prodigy who had created an automatic chess player that 'had attracted the awe-struck attention of the civilised world by the simplicity and daring of its mechanism', a whist player 'still more simply and boldly conceived' and 'a phonograph so perfectly constructed that people began to think that even Edison must soon begin to look to his laurels, or he would be eclipsed by the rising fame of this young man of thirty'. Challenged to go one better, he proceeded to build his lady automaton – a young lady indistinguishable in appearance, speech and manner from the real thing. Unfortunately, it transpired that the simulation was too good. Two suitors fell in love with her and asked her to marry them. Since she was designed to please, she agreed to both proposals. Things ended badly when one suitor stabbed the other along with his bride in the middle of the ceremony. In Alice Fuller's tale of 'A Wife Manufactured to Order', the man who had purchased an automaton wife soon grew to hate her imperfect perfection.[50] In many of these stories, the moral was one of discomfort with the blurring of the boundary between human and machine intelligence. Even Mitchell's tale about Baron Savitch ended with the baron's murder by the narrator, who knew his secret and was disgusted at the news that the machine planned to marry.

The presence of the phonograph in all these stories is quite striking. Invented by Edison in 1877, it was the breakthrough that really made him famous. In the patent he submitted in December that year, Edison described it as a device 'to record in permanent characters the human voice and other sounds, from which characters such sounds may be reproduced and rendered audible again at a future time'.[51] It was a device that stored aural information in a way that allowed it to be retrieved later – in Edison's original phonograph a needle recorded the sound on a cylinder covered in tinfoil. Tellingly, Edison later marketed a phonographic doll. The phonograph played a key role in another piece of contemporary fiction too – *Dracula*, which was published in 1897. In Bram Stoker's novel, the device was central to the protagonist's efforts to use the latest science to defeat an ancient enemy. 'He is experimenting, and doing it well,' Van Helsing warned his fellow vampire hunters of their prey. Responding with an experiment of their own, they used Mina's phonograph to keep a systematic record of their plans and discussions. For them, it was a machine that used the latest technology to make their words permanent. It meant they could consult the records of their past deliberations at any time.

Speculation about what automata, calculators and recording devices might achieve was not always fictional though. Automata were central to Nikola Tesla's vision of an electrical future laid out in his speculations on 'The Problem of Increasing Human Energy'. The piece in *Century Magazine* was part of Tesla's campaign to find investors to fund his grand dreams of electrical energy. The art of telautomatics, as he called it, would mean the end of war by making humans redundant – 'men must be dispensed with: machine must fight machine'. His goal was 'a machine capable of acting as though it were part of a human being – no mere mechanical contrivance, comprising levers, screws, wheels, clutches, and nothing more, but a machine embodying a higher principle, which will enable it to perform its duties as though it had intelligence, experience, reason, judgement, a mind!'[52] Tesla had no doubt that he could make such a machine. Simply by imagining himself as an automaton, a state he had thought about from an early age, he said, Tesla had already 'conceived the idea of constructing an automaton which would mechanically represent me, and which would respond, as I do

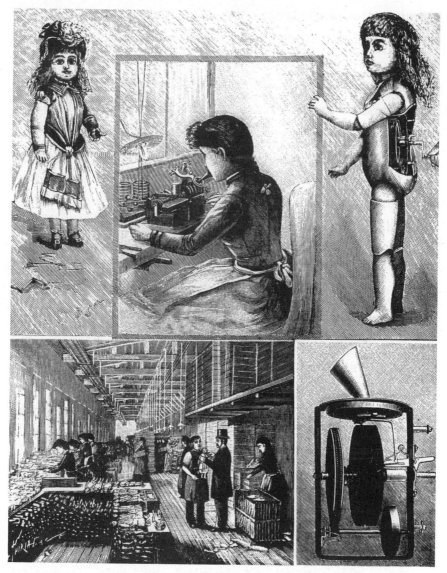

Manufacturing phonographic dolls.
De Natuur, 1894, 26

myself, but, of course, in a much more primitive manner, to external influences.'[53] He was the machine, and the machine would be him.

The machine would have to fulfil certain conditions. It would need to have 'motive power, organs for locomotion, directive organs and one or more sensitive organs so adapted as to be excited by external stimuli'. It would need to 'perform its movements in the manner of a living being'. It would need to have 'an element corresponding to the mind, which would effect the control of all its movements and operations, and cause it to act, in any unforeseen case that might present itself, with knowledge, reason, judgement, and experience'. That was where Tesla's dream of wireless energy came into play. Through it, automata could be controlled from a distance. Machines like these would have 'borrowed minds': 'the knowledge, experience, judgement – the mind, so to speak – of the distant operator' was 'embodied in that machine, which was thus enabled to move and to perform all its operations with reason and intelligence'. It would be 'like a blindfolded person obeying directions received through the ear'.[54] A few years earlier, Tesla has announced and demonstrated a model of a torpedo that could be operated wirelessly. He had even promised to put one on show at the Paris *Exposition Universelle* in 1900, wirelessly controlled from the other side of the Atlantic. He speculated that flying machines would make ideal automata too.

Speculations like these flourished in the culture of calculation that prevailed by the end of the nineteenth century. It was a way of thinking about the process of thinking that had a history stretching back to Babbage's efforts to find ways of calculating by steam, and even earlier to his youthful encounters with automata. Von Kempelen's chess-playing Turk (fake as it was) seemed to show that even playing chess could be carried out by machine – and a chess-playing machine duly featured in E.E. Kellett's tale to show off its protagonist's talent at creating thinking machines. The future that Babbage imagined as he laboured on his engines was a future in which calculation would be carried out on an industrial scale – and, as with other varieties of industrial work, the future would be organised according to the division of labour, with human computers replaced by mechanical ones at every opportunity. A working Difference Engine was never built, but the counting and calculating machines that Scheutz and others made still seemed to embody its potential. More than that, Babbage's Analytical Engine sparked a fascination with intelligent machines and their possibilities. It was an improved and miniaturised Analytical Engine that was inside Baron Savitch's skull, after all.

Tesla's telautomatic fantasy depended on *Century Magazine's* readers – and his potential future financial backers – being able to make that leap from humans transmitting their thoughts wirelessly at a distance and intelligent machines communicating among themselves. That scientific romances were cheek by jowl with scientific speculation in the pages of popular magazines certainly made that leap easier. But above all, the leap depended on the fact that by the beginning of the twentieth century, thinking with machines was commonplace. Cash registers counted in the money and counted out the change. Bank clerks used mechanical computers to keep accounts in order. Actuaries calculated the complexities of risk by machine. In government buildings, bureaucrats kept tabs on the statistics of life and death with Hollerith Electric Tabulating Systems. Machine thinking was everywhere because machine thinking was efficient. Machines could be relied on to keep on churning out their calculations without error. Humans could miscalculate. Machines, it was supposed, would not – although in fact they could and did. If getting to the future by the most efficient route required nice, accurate calculation, then it seemed clear that machines were best suited for the task. By steam or electricity, the best calculations would be mechanical ones.

Chapter 8

Flying High

Visitors to the Yorkshire gentleman Sir George Cayley's estate at Brompton sometime in 1853 would have witnessed an astonishing sight. One of those amazed witnesses was Cayley's own great-granddaughter, and long afterwards she remembered vividly 'a large machine being started on the high side of the valley behind Brompton Hall where he lived, and the coachman being sent up in it, and it flew across the little valley, about 500 yards at most, and came down with a smash. What the motive power was I don't know, but I think the coachman was the moving element, and the result was his capsize and the rush of watchers to his rescue. He struggled up and said, "Please, Sir George, I wish to give notice. I was hired to drive and not to fly."'[1] It seems unlikely that the coachman (who was actually Cayley's groom, John Appleby) really was the source of motive power for this device. It is more probable that the contraption was pulled along the ground by horses until it gained enough lift to become airborne. The machine in question was likely the 'governable parachute' that Cayley described in the *Mechanics' Magazine* the previous year, remarking that 'safe descent and steerage of such

𝔐echanics' 𝔐agazine,

MUSEUM, REGISTER, JOURNAL, AND GAZETTE.

No. 1520.]   SATURDAY, SEPTEMBER 25, 1852.   [Price 3*d*., Stamped 4*d*.

Edited by J. C. Robertson, 166, Fleet-street.

SIR GEORGE CAYLEY'S GOVERNABLE PARACHUTES.

Fig. 2.

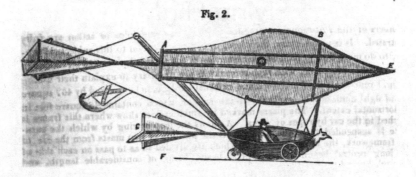

Fig. 1.

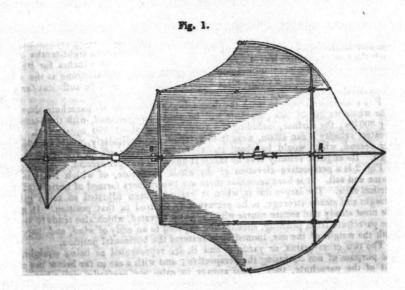

George Cayley's governable parachute.
Mechanics' Magazine, 1852

aerial vehicles has been abundantly proved, when properly adjusted, by their being launched from hill-tops into the valleys below'.[2] It was not Cayley's first attempt at flight, though – he had experimented a few years earlier using 'a boy of about ten years of age', who 'floated off the ground for several yards on descending a hill, and also for about the same space by some persons pulling the apparatus against a very slight breeze'.[3]

Cayley had been fascinated by flight for quite some time. As a baronet and one of the landed gentry, he cut a curious figure for an engineer and inventor. The sixth Baronet of Brompton – he inherited the title and the estate at Brompton Hall from his father in 1792 – was a man of means and leisure. He was not the kind of man who might normally be expected to be interested in getting his hands dirty. He had, however, been tutored by the mathematician and fellow of the Royal Society George Walker, as well as the Welsh electrical experimenter George Cadogan Morgan, which must have helped spur his technological interests. He had the resources and the time to develop those interests, too. As an improving landlord, he was involved in a variety of schemes to drain and reclaim land for agriculture. He was a reformer in politics, too, and briefly the MP for Scarborough after the passing of the Great Reform Act in 1832. Significantly, he had been financially involved in helping to establish both the National Gallery of Practical Science and the Royal Polytechnic Institution.

Cayley had been thinking seriously about the possibilities of powered flight as early as 1800. Convinced as he was that 'Aerial Navigation will form a most prominent feature in the progress of civilisation during the succeeding century', he speculated about balloons that could be steered and flown through the air rather than being entirely at the mercy of currents, going whichever way the wind blew. He imagined balloons from which 'a long and narrow car or boat must be suspended', along with a rudder and a source of power. In 1820, he even carried out experiments with a model airship along those lines on his estate at Brompton. Cayley was not just fascinated by balloons, though. He wondered about the possibility of winged flying machines as well, and even studied birds in flight to understand the dynamics of their motion through the air. He was interested in the engines that would power these aircraft of the future as well. He was unconvinced that the steam engine could ever be made light enough, so he experimented with hot air engines instead. He made a number of attempts to establish a society for the encouragement of aeronautical invention.

Cayley was not the only one. During the early 1840s, two lacemakers, William Henson and John Stringfellow, tried to establish the Aerial Transit Company with a view to promoting the Aerial Steam Carriage they had patented in 1843. Both men were experienced engineers. Henson, for example, had acquired a patent for improvements in lace-making machinery in 1835. Their flying machine patent described 'certain improvements in locomotive apparatus, and machinery in conveying letters, goods, and passengers from place to place through the air'.[4] It all depended on a new and lighter design of steam engine that Henson had also patented. It was an immensely ambitious scheme. Henson and Stringfellow, along with others invested in the scheme, were determined to turn their fantasies into reality. However, their attempts to establish the Aerial Transit Company came to nothing in the end. Efforts to get the bill of incorporation through Parliament descended into farce. They recruited the newspaper publisher Frederick Marriott, who had been involved in both the *Morning Chronicle* and the *Illustrated London News*, to their cause. He commissioned a number of illustrated trade cards for the company, each displaying a different scene from a future in which the Aerial Steam Carriage could be seen flying over a variety of landscapes, both familiar and exotic. There it was,

Henson's Ariel above the Pyramids.
Lithograph by W.L. Walton (London: Ackermann & Co., 1843)

flying over London, or flying across the Nile with the pyramids in the background.

Henson and Stringfellow did build and experiment with scaled-down models of the Aerial Steam Carriage over the next few years, but their attempts to persuade investors that there really was a future in their invention was a failure. From the very beginning, more sober observers were unconvinced that the machine would ever fly. 'We were of the number of those who looked at the announcement, that Mr Henson had constructed a machine that could fly, with great incredulity,' said the *Mechanics' Magazine*. It was, however, an incredulity 'arising rather from a recollection of the many previous attempts of the like kind, and the miserable uniformity of their failure, than from any want of faith in the resources of science and art to achieve even greater triumphs'.[5] The magazine still devoted a front page to it, nevertheless, although that it was the 1 April edition may be significant. The *Illustrated London News* made space for it in their 1 April edition, too, and they were similarly unconvinced. They thought it might take off, 'but a few, a very few seconds, must ensure its descent to earth, despite the operation of the steam propellers'.[6] Henson tried to rally Cayley to their cause, but Cayley, despite (or maybe because of) his own experiments, thought balloons were a safer route to the future of flying.[7] 'I had thought that you had abandoned the subject,' he replied to Henson's plea for support, 'which ... you had rushed upon with far too great confidence as to its practice some years ago.'[8]

Balloon flight was relatively common by the 1840s, and flights over comparatively long distances had been attempted. In 1836, the balloonist Charles Green, for example, accompanied by another balloonist, Thomas Monck Mason, and other passengers, took to the air from London's Vauxhall Gardens and landed the following day in Weilburg, Nassau (now part of Germany). In 1838, accompanied by the Duke of Brunswick, Green ballooned across the Channel again, landing in Neufchâtel, France. The problem with balloon flight was that it was largely at the mercy of prevailing winds. Cayley, in a letter to the *Mechanics' Magazine* in 1837, offered suggestions as to how balloons might be steered and how to provide them with a motive power, and in 1840, Green exhibited a 'miniature balloon armed with screw-propellers driven by a spring', at the Royal Polytechnic Institution as he tried to canvass support for an even more ambitious Atlantic crossing.[9] No surprise, then, when the *New York Sun* announced on 13 April 1844 that an Atlantic crossing by balloon really had taken place. The successful aviator was Thomas Monck Mason himself, and the newspaper enthusiastically described how he had adapted some of Cayley's ideas in his efforts. It was, the *Sun* was sure, 'unquestionably the most stupendous, the most interesting, and the most important undertaking, ever accomplished or even attempted by man. What magnificent events may ensue, it would be useless now to think of determining.'[10] It was all a hoax, though, perpetrated by none other than Edgar Allan Poe.

The hoax's reception showed just how plausible – and how close – a future of flying seemed to many people already. It all depended, after all, on people being willing to believe that this really could happen. Balloons were soon being seen as weapons of war, as well – and the role of flight in warfare would be a preoccupation for military men and writers of scientific romance for the rest of the century. In 1849, the Austrian Army used incendiary balloons during the siege of Venice, though without much success. Observation balloons were deployed by both sides during the American Civil War to help direct artillery fire more effectively. The balloonist Thaddeus Lowe, an assiduous promoter of balloons for transatlantic travel, was appointed the Union Army's chief aeronaut by Abraham Lincoln. On 16 June 1861, he succeeded in sending a telegraph message from the balloon *Enterprise*, floating 500 feet above the ground, to Lincoln at the White House, to demonstrate the value of balloons in war. In 1862, two British Army officers, Captain F. Beaumont and Lieutenant George Grover, carried out ballooning experiments at Aldershot. It took until 1878 to convince the army, but a year later it bought its first balloon, the *Pioneer*. In 1890, the British Corps of Royal Engineers established a balloon section, along with their own factory and ballooning school.

Cayley had speculated that airships could be driven through the air by propeller. To allow them to slide more easily through the atmosphere he suggested that airship balloons should be elongated rather than spherical and calculated what the best proportions might be. In 1851, the Australian doctor, politician and former convict

William Bland patented his Atmotic Ship – an airship that would be powered by steam and would, Bland hoped, make the journey from Sydney to London in less than a week. The ship could carry passengers, the mail and other goods, Bland speculated. It offered a safer and faster way of travelling, as well as a way of exploring otherwise inaccessible terrain like the interior of Australia itself. Like almost everyone else who speculated about flying, Bland was alert to the military possibilities. 'Could or would this method of travelling tend to place the different races of man too much on a par with each other as to their aggressive powers?' he asked, and 'if so, would it not enable the barbarous nations of the earth by their vast aggregate numerical superiority, to overwhelm the civilised portions of the globe?'[11] While Bland was planning his airship, in France, the engineer Henri Giffard succeeded in actually flying his steam-powered, cigar-shaped and hydrogen-filled airship for 27 kilometres in 1852.

Just at the beginning of the American Civil War, Solomon Andrews flew his airship *Aereon* over New Jersey and tried to persuade Lincoln that it would be a valuable asset for the Union Army. It consisted of three cigar-shaped balloons, each 80 feet long, with a gondola attached beneath them, and could be steered with a rudder. A few years later, Frederick Marriott, Henson and Stringfellow's former partner, who had since emigrated to the United States, flew his model powered airship, the *Avitor*, over San Francisco in front of admiring crowds. Mark Twain thought the flight was 'bound to stir the pulses of any man one talks seriously to about, for in this age of inventive wonders all men have come to believe that in some genius' brain sleeps the solution of the grand problem of aerial navigation', and then, 'with railroads, steamers, the ocean telegraph, the air ship – with all these in motion and secured to us for all time, we shall have only one single wonder left to work at and pry into and worry about – namely, commerce, or at least telegraphic communion with the people of Jupiter and the Moon'.[12] In 1883, another French inventor, Gaston Tissandier, who had escaped the Prussian siege of Paris by balloon a decade or so earlier, flew an airship powered by a Siemens electric motor. In 1888, Friedrich Wölfert flew one with a petrol engine designed by Daimler. The sky was being colonised, and a future in which airships dominated the skies was looking daily more plausible.

Inventors still hoped that they might solve the problem of powered, heavier-than-air flying as well. In 1868, the Aeronautical Society of Great Britain, founded just over a year earlier in 1866, to promote experiments in flying, held an exhibition at the Crystal Palace in London. The prize on offer even tempted John Stringfellow to compete. The model flying machine he entered for exhibition was steam-powered and notable for its lightness. He won the £100 prize for his model steam engine, which weighed only 16 pounds. Newspaper commentators thought the exhibition showed that 'the *possibility* of aerial locomotion by flying machines capable of sustaining great weight will be admitted by most persons who will frankly consider such facts as have now been adduced'. Now, 'men of science have at last given their attention to a problem which can only be solved or proved insoluble by co-operation, and by a strict attention to experimental methods'.[13] Stringfellow even had the opportunity to show off his machine before the Prince of Wales and his brother the Duke of Edinburgh at a grand fete organised at the Crystal Palace, who saw how the machine 'travelled across the transept and a considerable distance down the nave'. During the flight, 'the carriage had an evident tendency to rise in the air, as was seen by the fact that the line on which it was suspended and along which the pulley travelled was raised several feet'.[14]

A few years later in 1875, Thomas Moy tried out his Aerial Steamer at the Crystal Palace. This was an ambitious experiment carried out at full scale. The steamer was fourteen feet long and driven by a steam engine turning two 'aerial propelling wheels'. The whole contraption weighed 216 pounds. It was tried out on a circular track 30 feet in diameter around one of the Crystal Palace's ornamental fountains. Moy thought that the machine would need to travel at 35 miles per hour to get off the ground. It only managed twelve. In France, the electrical engineer Clément Ader started experimenting with flying machines in the 1880s. His first attempt, the *Ader Éole* had wings like a bat, a wingspan of 46 feet, a steam engine weighing 112 pounds and a total weight of 660 pounds. In the *Éole*, Ader managed to take off and hop along for about 160 feet, rising to maybe eight inches from the ground. In his next attempt in the *Avion II*, he later claimed to have flown for about 330 feet, but this was never verified. His experiments did come to the attention of the French War Office, nevertheless. They financed the building of the *Avion III*, which never quite succeeded in getting off the ground during experiments in 1897.

But attempts to get flying machines into the air were by now relentless. During the early 1880s, John J. Montgomery, later a professor at Santa Clara University in San Francisco, took a number of wild rides in a succession of unpowered flying machines, flying distances of up to 600 feet. Like many flying enthusiasts, he spent time studying the dynamics of bird flight in search of inspiration. Cayley devoted a great deal of attention to the ways in which birds held their wings while flying, for example, observing that 'when large birds, that have a considerable extent of wing compared with their weight, have acquired their full velocity, it may frequently be observed, that they extend their wings, and without waving them, continue to skim for some time in a horizontal path'. He carried out careful calculations of how the angle at which the wing was held supported flight in different circumstances.[15] Lord Rayleigh was another individual who devoted attention to bird flight in the interests of aeronautics. He had been speculating about the ways birds flew since the 1880s, and he brought the full weight of Cambridge mathematics and precision physics to bear on the question. In 1889, he was involved in a lively dispute in the correspondence pages of *Nature* about the flight patterns of albatrosses. His researches culminated in his Wilde Lecture (named after the electrical engineer Henry Wilde) to the Manchester Literary and Philosophical Society in 1900 on the mechanical principles of flight.[16] In 1909, his former assistant at the Cavendish Laboratory, Richard Glazebrook, who at Rayleigh's recommendation had been made director of the National Physical Laboratory, invited him to head its Advisory Committee on Aeronautics.[17]

Sir Hiram Maxim, inventor of the machine gun, almost succeeded in getting a huge, steam-powered flying machine into the air in 1894, before going on to develop tethered flying machines for fairgrounds. The astronomer Samuel Pierpont Langley experimented with power model machines on the Potomac River. Rudyard Kipling called them 'a marvel of delicate craftsmanship', and one of them flew a distance of more than 4,000 feet. He caught the attention of the assistant secretary to the US Navy, Theodore Roosevelt, who set up a committee to examine the experiments. In 1898, Augustus Moore Herring, who had assisted Langley in his experiments, claimed to have flown for a distance of 70 feet along a Michigan beach. A year later, Gustave Whitehead claimed to have flown a distance of 1,600 feet in his steam-powered flying machine – but as in Herring's case there were no witnesses. In England, in the same year, Percy Pilcher planned a powered flight in his machine at Stanford Hall, but died demonstrating his unpowered flying machine, *Hawk*, before the experiment could take place.

By 1899, the two brothers, Orville and Wilbur Wright, had started their own experiments in flying. They were both experienced mechanics, having run a bicycle shop in Dayton, Ohio, since 1892. After deciding to try their hands at flight, the brothers went about it systematically, contacting the Smithsonian Institution in Washington DC with a request for an exhaustive reading list. They also wrote to Octave Chanute, author of the voluminous *Progress in Flying Machines* with its meticulous descriptions of 'how much has been accomplished toward overcoming the various difficulties involved, and how far the elements of a possible future success have accumulated within the last five years'.[18] Among other things, Chanute advised them to find a good spot for experiment, preferably a beach with a good breeze and soft landing. The advice led them to Kitty Hawk in North Carolina, where over the next few years they focused their attention on controlling flight in their unpowered prototype machines. Successful flight, they decided, depended on the pilot making themselves familiar with the apparatus, adjusting themselves to its idiosyncrasies – just like riding a bike, in fact. It was not until 1903 that they began to experiment with powered flight, trying out various kinds of engine. And then, at the end of the year, they flew.

The *Wright Flyer* was built of spruce and ash and was powered by a lightweight, four-cylinder, petrol-fuelled internal combustion engine, designed and built by their assistant, Charlie Taylor, connected to the two wooden propellers by a bicycle chain. Their first serious attempt, on 14 December 1903, was a failure, and they needed three days to repair the damage caused by the crash landing. On 17 December, they tried again, with Orville at the helm, and he flew for about twelve seconds, a distance of about 120 feet. They flew several more times during the rest of the day, the two brothers taking turns in the flying seat, until a final crash damaged the machine beyond immediate repair. They never flew that particular machine again, but over the next few months the brothers flew many more flights in improved machines, going further and further each time. By 1905, they could fly for several miles in the *Wright Flyer III*, in circles as well as in a straight line. Ohio newspapers crowed that other aeronauts and balloonists were 'in eclipse as air navigators since the advent of Wilbur and Orville Wright, who have apparently solved the problem of aerial flight without the aid of gas bags'.[19]

The Wrights' achievement did not make national and international headlines at first, largely because the brothers were reluctant to say too much until they had a practical machine ready for the market. Reports that dripped out of Kitty Hawk were treated with a mixture of suspicion and incredulity on both sides of the Atlantic. Only in 1908, with contracts signed for commercial production both in Europe and the United States, did they start on a round of public demonstrations. By then, others were already hot on their tail. In France, for example, Louis Blériot and his Recherches Aéronautiques Louis Blériot, developed a series of Blériot machines capable of flying for at least short distances. By the beginning of 1908, Blériot was flying the latest version of his machine for miles across country. In 1908, the English newspaper the *Daily Mail* announced a competition and prize of £500 for the first flight across the English Channel and increased it the following year to £1,000. Following a number of unsuccessful attempts by other hopeful fliers, Blériot made his own attempt on 25 July 1909, leaving Calais just after 4.40am. Just 36.5 minutes later, he landed at Dover. Within barely a decade of the Wright brothers' first tentative flights at Kitty Hawk, flying machines were a practical, commercial – and military – reality.

Dreadnoughts of the air

Fascination with the possibilities of aerial warfare grew alongside the burgeoning fascination with the flying machines themselves. Balloons had been deployed as weapons of war – usually for observation

– since the middle of the century. John Edward Capper, a senior British military engineering officer who had served on the north-west frontier in India and in South Africa during the Boer War, and who was by 1903 the commander of the army's Balloon Sections at Aldershot, was an early visitor to the Wright brothers. Capper played a key role in developing British aerial power – he even piloted the army's first airship, the *Nulli Secundus*, as it flew over London in 1907. He was impressed and fully understood the military potential of flying machines. But what would aerial warfare look like? Faced with the prospect of flying machines that really flew, Lord Northcliffe, the *Daily Mail*'s proprietor, thundered 'that England is no longer an island'. Now, the 'aerial chariots of a foe will descend on British soil if war comes!'[20] And for many, there was no 'if' about the prospect of future war. It was widely regarded as a certainty. Writers of scientific romance made much of the prospect – and aerial warfare played a key role in their renditions of the war to come.

To many, naval warfare seemed a plausible model for war in the air. By the beginning of the Edwardian era, Britain's dreadnoughts were the last word in technological warfare. Those seemingly impregnable ironclads provided the model for what warships of the air might look like. While writers of scientific romance fantasised about what would happen when flying machines ruled the skies, more hard-headed men were working to make sure that Great Britain continued to rule the waves. Navies were transformed during the second half of the century as the other European powers (and, increasingly, the United States) tried to challenge British supremacy at sea and the British innovated to maintain their supremacy. By the end of the century, battleships were metal behemoths, carrying cannon that could wage war at a distance. The first ironclad battle-ship – the *Gloire* – had been built in France to try and match the British Navy. The British did not take long to catch up, however.

Their ironclads soon outclassed their French counterparts. Ironclads played a key role in the American Civil War. Powered by steam, these new battleships looked less and less like the traditional fighting vessels that had ruled the waves in the past. Their iron- (and later, steel-) clad hulls lay close to the waterline, and they carried fewer, but far more powerful, armaments. As well as guns, ironclads were sometimes equipped with rams that would allow them to sink enemy vessels by holing them below the waterline.

However much the authors of scientific romances might credit the invention of their fictional dreadnoughts of the air to impecunious inventors wasting away in garrets, the reality of building and designing the ironclads that ruled the waves was rather different. Shipbuilding depended on committee men and batteries of experts. Back at the beginning of the century, Marc Brunel might have gained the bulk of the credit for the block-making machinery at the Portsmouth dockyards that had revolutionised their production and the capacity of the British Navy to stay at sea, but the machines needed Henry Maudslay and his mechanics to actually make them. Similarly, Isambard Kingdom Brunel's leviathan *Great Eastern* might have been entirely his in the popular imagination, but it took armies of engineers to get it into the water. In just this way, the American rush to build ironclad ships by both sides in the Civil War meant committee work and careful consideration of different designs by battalions of naval engineers.[21] In this the Americans were following models of naval innovation already established in Britain and France. Ships were designed by teams of architects, their design exhaustively – and sometimes bitterly – scrutinised and debated before they got off the drawing board.

Sir Edward Reed, who, as the Royal Navy's chief constructor during the 1860s was one of the main scrutineers of such designs, remarked that 'the efficiency of its ironclad fleet is of foremost importance to a small, isolated, maritime country like this, anchored on the edge of a continent like Europe, entrusted with the care of world-wide interests, and charged to maintain its power upon the sea at a time when the spirit of invention is setting at naught all past systems of ocean warfare, and mocking at every trace and tradition of the times when we won our naval renown'. The new breed of ship was more powerful – and needed to be more powerful – in every way. As Reed argued, 'enormous modern steam-engines', represented 'an entirely new field for the experience of our officers and men, and a field which it is absolutely necessary to cultivate with the greatest assiduity and care, as it is by such engines that the great Channel Fleets of England will be propelled for many years to come'. It was better 'to see our floating Channel and Mediterranean Fortresses well armoured, well armed, and well supplied with steam propellers', even if it meant the new behemoths were less agile and manoeuvrable than the sail-powered warships of the past, and he looked 'with lively expectation to the production of much more powerful guns than we have yet seen.'[22]

By the 1890s, ironclad battleships typically carried four big guns for blasting enemy ships at a distance, as well as smaller guns designed to be used when the ship moved in for the final kill. Ships were getting bigger, as well. The Royal Sovereign class of ships were 410 feet long, powered by two huge steam engines, each driving a propeller shaft. They could travel at speeds of up to twenty miles an hour. Even they were not formidable enough, though. In 1905, the Admiralty ordered the building of HMS *Dreadnought*, the first of an entirely new class of warship. These new ships were bigger, faster, better armed and better defended – practically impregnable floating fortresses. Susceptibility to torpedo attack was their only weakness. It was against ships like these that Nikola Tesla imagined his wireless torpedoes being deployed. In November 1898, he secured a patent for his plan to destroy battleships with torpedoes controlled by wireless by distant operators. His invention would revolutionise warfare, Tesla claimed. The ironclad behemoths that controlled the seas would be helpless against his weapon. The device would 'work a revolution of the politics of the whole world', and would 'make war so terrible, as well as expensive, as to make it prohibitory, and thus to assure peace between all the nations'.

Tesla's predictions might have been fantastic, but they were based on a reality in which these floating fortresses of steel dominated naval warfare with weaponry that challenged the balance of power. H.G. Wells even imagined the consequences of land ironclads dominating terrestrial battlefields in the same way.[23] Guides were published to help ships' captains identify the colossuses against which their own vessels would be pitted. In 1898, for example, Frederick Jane published the first edition of *All the World's Fighting Ships*, illustrated with silhouettes of each battleship, divided by country and listing details of their armaments and capabilities.[24] Jane eked a living as a journalist and illustrator during the 1890s. He had made his name in 1889 when commissioned by *Pictorial World* magazine to cover naval manoeuvres taking place at Spithead, giving him an opportunity to hone his skills by producing sketches of more than 100 battleships. This information was fast becoming necessary knowledge. Jane also produced a series of sketches for the *Pall Mall Gazette*, entitled 'Guesses at Futurity', picturing life in the year 2000. They featured streets illuminated by electric lights suspended from airships, and interplanetary flight in an electromagnetically powered ship. In 1909, he added *All the World's Airships* to his list of publications.

Jane was a writer of scientific romances, as well as an illustrator. One of the more prolific authors whose books Jane illustrated was George Griffith. From a career as a struggling schoolmaster and forays into journalism, Griffith made a name for himself in 1893 when *Pearson's Magazine* serialised his novel, *The Angel of the Revolution*. Griffith gave free rein to his radical political inclinations, describing how a determined group of anarchists succeeded in overturning the European balance of power, ushering in a new age, thanks to their domination of the air. Their airship, invented by a penniless genius in his garret, terrorised Europe, destroying fortresses and reducing cities to rubble at a whim. His 1895 novel, *The Outlaws of the Air*, also serialised in *Pearson's*, pitted radical against radical for control of the skies. A group of peaceful anarchists, funded by a benevolent millionaire, had set out to live in their own island utopia in the Pacific, only to have their secret airship technology stolen by a violent sect, set on destruction. The pacifists had no choice but to enter into battle themselves, their aerial technology deployed to protect the world from terror delivered from the sky.

An aerial battle from George Griffith, *The Outlaws of the Air* (London: Tower, 1895).

The flying behemoths that Griffith described really were dreadnoughts in the sky – and that was just how Frederick Jane portrayed them. The *Ariel* was 'a long, narrow, grey-painted vessel almost exactly like a sea-going ship, save for the fact that she had no funnel, and that her three masts, instead of yards, each carried a horizontal fan-wheel, while from each of her sides projected, level with the deck, a plane twice the width of the deck and nearly as long as the vessel herself'.[25] The ship was powered by 'four somewhat insignificant-looking engines', of which 'one drove the stern propeller, one the side propellers, and two the fan-wheels on the masts'. The armament consisted of four long, slender cannon, two pointing over the bows, and two over the stern.'[26] They were massive airguns that fired missiles of immense destructive power. When the *Ariel* attacked Kronstadt during its first battle, a single missile reduced the fort to nothing: 'The fort seemed to crumble up and dissolve into fragments, and a few moments later a dull report floated up into the sky mingled, as he thought, with screams of human agony.'[27] Griffith offered a graphic example of Lord Northcliffe's fear that uncontested air power was absolute power – and it was a fear widely shared at the end of the Victorian century.

But what about the projection of air power beyond the confines of the Earth itself? Jane's illustrations of dreadnoughts in the sky helped make Griffith's stories real for their readers by reminding them that the airships flying through them were not that different from the seabound battleships that really did determine the fate of empires. Scientific romances of flying machines extrapolated from the realities of naval warfare and the determined efforts of ambitious inventors to get their contraptions off the ground. Other authors had the players in their romances crossing unpassable mountains, deserts or seas of ice to discover lost lands and hidden civilisations. The relationship between aerial battles and adventurous explorations through the skies and real sea battles and naval expeditions must have been clear to those who wrote and read those tales. If nothing else, the romances appeared in the same monthly magazines that might also feature essays about the latest science of naval warfare or journeys to the diminishing number of unexplored regions left on Earth. Adventures that took their heroes even further afield were of the same kind. Victorian inventors of the future, whether they were writing scientific romances or struggling to get their machines into the air, ultimately had their sights set on higher things.

Landscapes in the sky

The first (known) photograph of the Moon was taken on 26 March 1840 by John William Draper, the recently appointed professor of chemistry at New York University. The English-born Draper would later go on to write the highly divisive and influential *History of the Conflict between Religion and Science*. The photograph he took was a daguerreotype – the process developed by Louis Daguerre and announced only the previous year. It was a blurred image, but it inaugurated a tradition of turning this latest technology to the sky and seeing other worlds in new ways. Within a decade, there were far better images to be admired. In 1849, John Adams Whipple and William Cranch Bond used the Harvard College Observatory's telescope to capture an impressive daguerreotype image. Another of Whipple's Moon daguerreotypes was exhibited at the Great Exhibition in London in 1851. Admiring Bond's daguerreotype at the Crystal Palace, James Glaisher commented that 'a process by which transient actions are rendered permanent, and which enables *nature* to do her own work, or, in other words, which causes facts permanently to record themselves – is so well fitted for the purpose of science to be long overlooked'.[28] Others followed. Warren De la Rue, inspired by Whipple's success, turned his attention to lunar photography. A few years later, Thereza Dillwyn Llewellyn, working with her father at his Welsh estate at Penllergaer succeeded in taking a number of pictures.

De la Rue, making full use of his private laboratory and observatory at Cranford, soon went even further. He had the time and the money to indulge his passion for chemistry and astronomy, and lunar photography combined both of them. His photographs were so sharp that they showed 'details on the Moon's surface sufficiently clear to admit of delineation under a microscope provided with a camera lucida, and thereby furnishing materials for a more accurate selenography than has heretofore existed'.[29] Even more astonishingly, he succeeded in taking stereoscopic pictures of the Moon, too, which meant that viewers could see the images in three dimensions. Awarding De la Rue the society's gold medal for his achievement, the Royal Astronomical Society's president marvelled that 'Mr De La Rue has brought to light details of dykes, and terraces, and furrows, and undulations on the lunar surface, and of which no certain knowledge had previously existed'.[30] De la Rue sent copies to the now venerable old astronomer John Herschel (himself the inventor of the word photography), who marvelled that 'the view is such as would be seen by a giant with eyes thousands of miles apart: after all, the stereoscope affords such a view as we should get if we possessed a perfect model of the moon and placed it at a suitable distance from the eyes, and we may be well satisfied to possess such means of extending our knowledge respecting the moon, by thus availing ourselves of the giant eyes of science'.[31]

Herschel was familiar enough with speculation about the lunar landscape. His father, William, discoverer of the new planet Uranus and, along with his sister Caroline, an assiduous cataloguer of nebulae, had also surveyed the Moon's surface with his telescopes. In communications to the Royal Society, William vividly described his observations of the lunar mountains and volcanoes he had glimpsed.

'It may, perhaps,' he told them, 'be esteemed to be a mere matter of curiosity to search after the height of the lunar mountains.' But 'when we consider that the knowledge of the construction of the Moon leads us insensibly to several consequences, which might not appear at first; such as the great probability, not to lay almost absolute certainty, of her being inhabited, we shall soon agree, that these researches are far from being trifling.'[32] In 1787, Herschel announced that he had seen an active volcanic eruption on the Moon. More strikingly, he had also seen clear evidence that there really was life – and intelligent life at that – inhabiting the lunar surface. He saw towns and cities on the Moon's surface, as well as roads, along with evidence of cultivation. Unsurprisingly, when the *New York Sun* perpetrated another scientific hoax in 1835 – announcing this time that inhabitants had been discovered on the Moon – they named John Herschel as the discoverer.

Just as with their other hoaxes, the Moon hoax worked because it was, in fact, perfectly plausible. The belief that other worlds were inhabited was a common one throughout the nineteenth century. Our own world, as natural theologians such as William Paley and Thomas Dick, argued, were evidence not just of design, but of the benevolence and wisdom of the creator – because our world had been created for us. In the same way, the other worlds that astronomy investigated must have been created for some purpose, too, as expressions of divine benevolence – and just as our world was created for us, they must have been created for their own inhabitants. Thomas Dick argued that 'new systems' were perpetually 'gradually forming in the distant regions of the universe.'[33] It stood to reason that the worlds those new systems contained were inhabited. As they were formed by physical forces acting as God had ordained, 'we must necessarily admit that a *direct interference of the Deity is necessary before such worlds*, after being organised, *can be replenished with inhabitants*'.[34] The blasphemous, bestselling (and originally anonymous) *Vestiges of the Natural History of Creation* argued of other worlds that 'every one of these numberless globes is either a theatre of organic being, or in the way of becoming so'.[35]

When William Whewell, shocked by the seeming blasphemy of the anonymous author of *Vestiges of the Natural History of Creation* tried to argue otherwise, his arguments were met with a combination of scorn and incredulity, even by the orthodox, so obvious was it that there must be inhabitants on other worlds. Whewell argued that 'dim as the light is which science throws upon creation, it gives us reason to believe that the placing of man upon the earth (including his creation) was a supernatural event, an exception to the laws of nature' – and he was clear that it was a supernatural event that could only have occurred once, whatever Dick said.[36] One reviewer in response mocked that 'we scarcely expected that in the middle of the nineteenth century, a serious attempt would be made to restore the exploded idea of man's supremacy over all other creatures in the universe; and still less that such an attempt would be made by one whose mind was stored with scientific truths'.[37] David Brewster even accused Whewell of degrading both astronomy and religion – he was just as misguided as the author of *Vestiges*, as far as Brewster was concerned. John Herschel simply laughed: 'So *this* then is the best of all possible worlds – the *ne plus ultra* between which and the 7th heaven there is nothing intermediate. Oh dear! Oh dear!'[38]

By the 1860s, the landscapes of the Moon were increasingly familiar to more and more people. It was clear, too, that this familiar landscape was probably inhabited. The same was happening to Mars, as well. In 1865, the Reverend William Rutter Dawes presented the Royal Astronomical Society with a series of remarkable and detailed drawings he had made of the Martian landscape over a number of years. He pointed to the 'long narrow strait running N.E. and S.W. in the northern hemisphere', and the 'curious forked shape; – giving the impression of two very wide mouths of a river, which however I could never trace'. He speculated that it 'might be that the sea has receded from that part of the coast, and left the tongue of land exposed'.[39] Warren De la Rue thought that Dawes' sketches were 'quite sufficient to construct a globe of *Mars*'.[40] Another Reverend, Thomas William Webb, remarked a few years later that 'we are ready to claim that globe as a close relation of our own, inferior indeed in magnitude and importance, if importance is indicated by an attendant, but arranged in a corresponding manner by the Great Creator as the seat of life and intelligence.'

In 1877, the Italian astronomer, Giovanni Schiaparelli, director of the Brera Observatory in Milan, announced that he could see networks of canals on Mars, criss-crossing the planet. Schiaparelli had a reputation as a cautious and reliable observer. If the canals he reported were really there, it seemed very unlikely that they could be entirely natural formations. But despite their apparently huge dimensions, they were difficult to see and many astronomers were dubious about their existence. American astronomer Percival Lowell, on the other hand, was quite sure he could see them. He had trained the powerful telescopes of his Flagstaff Observatory in Arizona on the planet and drew meticulous maps of the canals he saw on its surface. Mars possessed an atmosphere and clouds; there were signs of water and drought; and, above all, there were canals – a solution, Lowell suggested, to the Martians' overwhelming 'water question'. The planet's 'great continental areas' were all 'traversed by a network of fine, straight, dark lines'. They were clearly artificial – there was 'nothing haphazard in the look of any of them' and their 'most instantly conspicuous characteristic is this hopeless lack of happy irregularity'.[41] It seemed as if 'the whole of the great reddish-ochre portions of the planet is cut up into a series of spherical triangles of all possible sizes and shapes.'[42]

The Solar System's other planets were starting to look like distinct and different worlds as well. It was clear that Venus had an atmosphere and that most of the surface was covered by clouds, although at least one observer thought they had seen a 'lofty group of mountains penetrating the vapour-stratum supposed to form the greater part of the visible disc'. Photographs showed that 'in figure, Venus very closely resembles our earth', and astronomers speculated that beneath the clouds there were huge oceans.[43] Jupiter and its moons presented observers with 'one of the most diversified scenes in the heavens' as the 'miniature planets are seen overtaking, passing, meeting, hiding, and receding from one another in most beautiful and endless mazes'.[44] The planet's visible surface featured a series of belts between the poles and the equator, with 'dusky loops or festoons, whose elliptical interiors, arranged lengthways, and sometimes with great regularity, have the aspect of a girdle of luminous egg-shaped clouds surrounding the globe'.[45] Saturn had its spectacular rings, of course, and, like Jupiter, 'a large proportion of his bulky globe, 71,000 miles in diameter, is composed of heated vapours, kept in active and agitated circulation by the process of cooling'.[46] Even the further planets – Uranus and Neptune – had definite characteristics to be seen and studied.

Beyond the Solar System itself, new techniques like spectroscopy – breaking the light from distant stars down into its constituent parts – offered new knowledge about the nature of the Universe in its furthest reaches. Warren De la Rue turned his observatory into a laboratory, with batteries, induction coils and specimens of pure metals arranged alongside telescopes. The idea was to compare the light from the stars with the lights emitted by the combustion of those metals to try and identify the elements that made up those stars. As James Clerk Maxwell, the professor of experimental physics at Cambridge, put it, work like this showed that the distant stars were really very similar to the Sun and its Solar System. They were made from the same sort of stuff: 'When a molecule of hydrogen vibrates in the dog-star, the medium receives the impulse of these vibrations; and carrying them in its bosom for three years, delivers them, in due course, regular order and full tale into the spectroscope of Mr Huggins at Tulse Hill.'[47] New discoveries confirmed that the Universe was the same everywhere. They opened up a vista of distant stars as suns with their own solar systems orbiting around them, just as the Earth and the other planets orbited around our Sun.

The Moon, Mars, even Venus, Jupiter or Saturn were beginning to look more like places rather than mere lights in the sky. They had discernible features both similar and different to the familiar geography of the Earth. They had their own landscapes and skyscapes. Even the more distant corners of the Universe were starting to take on an air of familiarity. They could be seen – if only in the mind's eye – as places not that different from our planet. They were beginning to look like places – and places that might be visited. To some Victorian eyes, these landscapes in the sky might even look like territories waiting to be conquered. The Moon and other planets offered writers of scientific romances new stages to play out their fantasies about the future. As they contemplated a future in which flying machines could take them to the Moon and beyond, scientific romancers had raw material for their future-making in the form of technologies that existed already. And as powered flight through the atmosphere seemed increasingly certain, the prospect of flying even further seemed more plausible as well. When H.G. Wells, for example, had his hostile Martians launch their attack in machines that had been hurled through the Earth by huge explosions on their own planet's surface, he was imagining a technology not too far removed from what was already within the grasp of Victorian inventors of the future.

And now to the stars

When H.G. Wells sent men to the Moon two years after *The War of the Worlds*, rather than sending them on their way with explosives, he propelled them there with a strange new material called cavorite. The material was the invention of the mysterious Mr Cavor, an expert on 'the ether', and 'tubes of force', and 'gravitational potential' and 'things like that'. Cavor had succeeded in manufacturing a material that was opaque to gravity. After dreaming of the huge wealth that the material would generate for them, Cavor and his partner hit upon the plan of using it as the means of propulsion for a spaceship. 'Imagine a sphere,' Cavor explained to Mr Bedford, the story's narrator, 'large enough to hold two people and their luggage. It will be made of steel lined with thick glass; it will contain a proper store of solidified air, concentrated food, water-distilling apparatus, and so forth. And enamelled, as it were, on the outer steel ...' 'Cavorite?' 'Yes.'[48] The idea was that if the gravity opaque material were placed between any object and the Earth, not only would that object be made weightless, but the air above it as well, so that the object – in this case the spaceship – would be hurled upwards at great speed.

Cavorite was one ingenious solution to the problem of propulsion. Just like Cayley, Henson, Stringfellow and the Wright brothers, writers of scientific romances had to imagine ways of getting their machines up into the sky. Importantly, they had to be within the boundaries of plausibility to convince readers. Wells, for example, had Cavor explain that since there were materials opaque to other energies, such as light, or electricity, there must be a material opaque to gravity, too. Ellsworth Douglass and Edwin Pallander (the pseudonym of Irish scientist Lancelot Francis Sanderson Bayly) solved the problem by having the Earth itself, rather than the spaceship, move, in their short story 'The Wheels of Dr Ginochio Gyves'. Their ship had the ability to anchor itself in the ether so that as the Earth moved, it left the spaceship behind it. Pallander had employed a similar device a few years earlier in his *Across the Zodiac*, in which the *Astrolabe*'s captain explained that 'if I cannot exactly construct a ship to fly through space like a meteor or aerolite, I can at least design one that will remain motionless, or nearly motionless, while the planets, together with their satellites, move to and fro, above it, beneath it, or around it'.[49]

The commonest means of propulsion through space was electricity in some form or another. Frederick Jane had included a spaceship – the *Magnetiscope* – that apparently worked through electromagnetism in his 'Glimpses of Futurity' series of illustrations for the *Pall Mall Gazette*. When Thomas Edison led a fleet of spaceships from the Earth to Mars to punish the Martians for their attempted invasion in Garrett P. Serviss' unauthorised sequel to *The War of the Worlds*, they ran on electricity too. What Edison had done, was 'to create an electrified particle which might be compared to one of the atoms composing the tail of a comet, although in reality it was a kind of car, of metal, weighing some hundreds of pounds and capable of bearing some thousands of pounds with it in flight. By producing, with the aid of the electrical generator contained in this car, an enormous charge of electricity, Mr Edison was able to counterbalance, and a trifle more than counterbalance, the attraction of the earth, and thus cause the car to fly off from the earth as an electrified pithball flies from the prime conductor'. It was barely plausible as a method of propulsion but it was presumably just enough to provide a glimmer of credibility to the adventure story.

Not all tales of future space flight were about conquest. The planets offered opportunities for exploration and adventure, too. John

Jacob Astor had the *Callisto*'s crew set off on a jaunt that took them across the entire Solar System. As far as these denizens of future Earth were concerned, space and its contents were there for the taking. Flying over the surface of Jupiter, they marvelled at 'towering and massive mountains, and along the southern border of the continent smoking volcanoes, while towards the west they saw forests, gently rolling plains, and table lands that would have satisfied a poet or set an agriculturalist's heart at rest. "How I should like to mine those hills for copper, or drain the swamps to the south!" exclaimed

The spaceship *Callisto* from John Jacob Astor, *A Journey in Other Worlds* (New York: D. Appleton, 1894).

Col. Bearwarden.'[50] They speculated that if 'Jupiter is passing through its Jurassic or Mesozoic period, there must be any amount of some kind of game'. The travellers were on their journey to other worlds because they were 'tired of being stuck to this cosmical speck'. Earth's possibilities were 'exhausted, and just as Greece became too small for the civilisation of the Greeks, and as reproduction is growth beyond the individual, so it seems to me that the future glory of the human race lies in exploring at least the solar system'.[51] This was imperial space flight imagined by one of the richest men in America.

George Griffith similarly imagined the Solar System as a playground for travellers from Earth. The honeymooning couple who were the *Astronef*'s passengers set out on a jaunt through the planets. On the Moon, they found the skeletal remains of ancient lunar giants as well as their degenerated descendants still struggling to survive beneath the surface. Approaching Mars, they worried that the Martians, too, would 'have passed the zenith of civilisation, and are dropping back into savagery, but still have the use of weapons and means of destruction which we perhaps, have no notion of, and are inclined to use them'.[52] They had no compunction about retaliating savagely when the Martian airships did attack. They were entranced by the innocent and childlike flying beings who populated Venus and communicated through song. They marvelled at the crystal cities of Jupiter and their beautiful inhabitants, while speculating about their origins: 'Survival of the fittest, I presume. These will be the descendants of the highest races of Ganymede – the people who conceived the idea of prolonging human life like this and were able to carry it out. The inferior races would either perish of starvation or become their servants. That's what will happen on Earth, and there is no reason why it shouldn't have happened here.'[53] On Saturn, they encountered hideous submarine monsters who were no match for the *Astrolef*'s weaponry.

Stories like these about interplanetary travel offered Victorian authors and readers opportunities to ponder the future as it played out on a different canvas from the usual, Earthbound one. The danger of racial degeneration was a theme that preoccupied H.G. Wells, for one. It formed a key element in his scientific romance of the future, *The Time Machine*, published in 1895, with its degenerated Eloi and Morlocks. In his 'Stories of Other Worlds' (published originally in *Pearson's Magazine*, and later in novel form as *A Honeymoon in Space*), Griffith was playing the same kind of game, with the future played out on the planets, instead. The inhabitants of the Moon and Mars were degenerating as their civilisations declined along with the habitability of their worlds. The Jovians practised eugenics to improve and preserve their race. It showed to what extent dreams of empire were intimately wound up with these dreams of conquering space. Griffith and Astor's spaceships certainly had more in common with the dreadnoughts that dominated terrestrial seas than with dirigibles or the one-man machines that the Wrights and others were building. These vessels were illustrated as having long, thin hulls, tapering to a point at bow and stern, with a long superstructure in the middle. As Griffith's couple's encounter with the Martian hostile natives demonstrated, space-travelling humans were armed to the teeth.

These imagined behemoths were certainly a long way from Henson and Stringfellow's Aerial Steam Carriage, or the governable parachute that carried George Cayley's terrified coachman across the valley at Brompton. But even those early experiments were carried out with imperial intent. Cayley thought it was 'a national disgrace, in these enlightened locomotive times, not to realize, by public subscription, the proper scientific experiments, necessarily too expensive for any private purse, which would secure to this country the glory of being the first to establish the dry navigation of the universal ocean of the terrestrial atmosphere'.[54] Those images that

Frederick Marriott produced for the Aerial Transit Company, showing the Aerial Steam Carriage flying over exotic terrain, were telling, too, of imperial as much as commercial ambition. They showed how conquering the air would, like the expansion of the telegraph network, make the world easier to navigate. The vocabulary of flight was one of conquest, after all, and it is striking that war – and the terror of aerial war – played such a prominent part in speculation about the possibilities of mechanical flight. That made controlling the air – like controlling the seas – very much an aim of empire.

Epilogue

There was an eerie silence aboard HMS *Dreadnought* as anxious eyes scanned the bright blue sky above the Indian Ocean. It was eight days since HMS *Victorious* had climbed majestically into the sky above the Deccan Plateau and disappeared into the azure haze. It was three days since the three intrepid selenauts had stepped cautiously down from the landing lighter *Deliverance's* steel doors onto the surface of the Moon, and two days since they had departed on their way back to Earth. Now the world waited. The King and his entourage were there on the *Dreadnought's* main deck, searching the sky above through their opera glasses. Even with the best Marconi apparatus, contact with the crew had been patchy at best once the *Victorious* had been launched into the skies. The mission's commanders on the ground knew that the selenauts had landed successfully. They knew that *Deliverance* had left the lunar surface at the prearranged time. Beyond that, they knew nothing. Had the little landing lighter survived the fiery journey through the Earth's atmosphere? 'They're there!' someone shouted, and everyone strained to look through the glare. There was a tiny black dot getting rapidly bigger and bigger. Soon, everyone could see the dark, cone-shaped craft, suspended from three billowing parachutes falling towards the ocean.

Everyone held their breath as it disappeared beneath the surface, and everyone cheered when it rose back again. *Deliverance* had returned.

Over the next few days, the exploits of the men on the Moon filled the newspapers. What had they found? What was it like to walk on the surface of a strange new world? There were no signs of life on the Moon, at least not in the vicinity of the landing site. The selenauts had seen nothing that might indicate the presence either now or in the distant past of some declining civilisation. There were no signs of water either. The Moon's surface was an arid, barren plain stretching out towards distant hills and mountains. The selenauts had brought samples of lunar rocks back with them, and they would be subject to the most detailed analysis by leading men of science searching for evidence of what the Moon might offer the Empire. Were there metals there to be mined? Were there valuable minerals? Plans were being made already for a second *Victorious*, and a third. If there were resources on the Moon to be exploited, then it was essential that the means should be found to do so. The selenauts might have found no water in the Sea of Tranquillity, but there were tantalising hints from spectroscopic surveys that some might be lurking somewhere on the surface. If there really was water there, then the Moon was ripe for colonisation.

Could this have happened? Could Victorian adventurers really have landed on the Moon? In reality, no, they couldn't have. They simply lacked the knowledge and the technology necessary for such a task. There never could have been moustachioed selenauts exploring the lunar surface. That is not the point, though. What the Victorians lacked in terms of knowledge and technology, they made up for in imagination. For them, going to the Moon, and beyond to the planets, was within the boundaries of the possible. In that sense, at least, the Victorians really did take us to the Moon. When *Apollo 11* took off from the Kennedy Space Center on 16 July 1969 – just 60 years after the scene imagined at the beginning of this book – and when the *Eagle* landed on the lunar surface on 20 July, it really was the culmination of a technological fantasy that began with the Victorians. What this book has tried to describe is the emergence during the course of the nineteenth century of new ways of thinking about and organising science that were directed at the future in a wholly new and unprecedented way, and some of the key consequences of that reorientation. It is, by and large, the way we think about and organise science now, and the book is also an invitation to think about what it means that we still do things the Victorian way.

I started by suggesting that the Victorians invented the future as we know it. During the nineteenth century, the future became a different place: a place that was both an extension of the present and wholly different from it. The Victorian future was built around progress. Their world was not static any longer; their universe was in a constant state of evolution, and just as nature evolved, Victorians thought that society should evolve as well. The Solar System had started out as a cloud of dust and gas, slowly coalescing into its present form through gravity. The spark of life had its origins in the primeval mud and had evolved through aeons to produce the Victorians themselves. Victorian society was on the way to somewhere, too. It would get there by reform rather than revolution, they told themselves. Progress was natural and built into the order of things. This meant that the future was a destination. The future would be different – though different in very specific ways that would preserve its identity as distinctively Victorian. Fantasies about going to the Moon and beyond, and efforts to turn that fantasy into reality, were products of this new way of thinking about the future.

Science was what would take the Victorians to the Moon, and science was what would take them into the future, as well. When men of science battled to reform their institutions at the beginning of the Victorian age, they were embarking on a wholesale reorganisation of the sciences. They were trying to carve out new roles for themselves, and new uses for their disciplines – and looking towards the future was integral to their efforts at making themselves useful in novel ways. Their science was made for future-making. And if this sounds unsurprising – what else should science be for? – then that should be a reminder that we have inherited this perspective about science and the future from the Victorians. That this was the place that science should occupy was not self-evident at all at the beginning of the Victorian century. Making it so took time and labour. People like Charles Babbage, or William Thomson, or James Clerk Maxwell, had to demonstrate to those around them that they could be trusted with the future – and so did people like H.G. Wells or George Griffith, for that matter. So, this is a story about creating a new kind of expert, as well as a story about creating a new kind of future.

Victorian science was built around the value of discipline – and disciplined men was exactly what the new breed of scientific experts were meant to be. Science was not for just anyone. To be a man of science was to have a particular kind of mind. That was what the men who set about reforming the Royal Society thought made them different from their predecessors. Sir Joseph Banks and his coterie simply did not have the kind of disciplined minds that were needed to produce disciplined knowledge. That was what Babbage in the 1820s, or William Robert Grove in the 1840s, thought reform was all about: the Royal Society needed to be reformed so that it could be seen to have the right kind of discipline to be useful. Faced with the popularity of fads like mesmerism and table-turning in the 1850s, Michael Faraday argued that it was because he and people like him had learned how to discipline their minds that they could be trusted to tell the difference between fact and fancy. That was what distinguished their expertise from the outpourings of unscrupulous charlatans. It was by persuading the university's authorities that the teaching of experimental physics – like the teaching of mathematics – was a good way of disciplining young minds that Maxwell convinced them that the Cavendish Laboratory belonged in Cambridge.

Self-discipline was – or was meant to be – the supreme masculine virtue for the Victorians, so it should be no surprise that their science came to embody it. In fact, it was by making their science the embodiment of that Victorian virtue that men of science made their discipline dominant. The possession of disciplined minds was what was supposed to be the difference between men and women. Men could be trusted to keep themselves under control while women were at the mercy of their uncontrollable bodies. That was why there were no women in Samuel Smiles' *Self-Help*, after all – and that was why so many of his examples were men of science, engineers and inventors. Men like that were what they were precisely because they possessed those virtues, and that was what made their science possible. This was why Faraday, for example, was celebrated as the exemplary man of science, and why there was such an emphasis on his being a self-made man. Faraday had made it from rags to scientific riches because he knew how to discipline himself. He had that characteristic cast of mind – and for Victorian men of science there could be no doubt that this was a masculine trait.

This is why showmanship was such an important element of the business of invention. At the Adelaide Gallery, or the Great Exhibition of 1851, or the grand industrial exhibitions of the second half of the century, the inventor was as much on show as their invention. Men of invention were – and were supposed to be – powerful and charismatic figures. That famous and iconic image of Isambard Kingdom

Brunel, posing in his stovepipe hat, cigar in mouth, in front of the *Great Eastern's* massive launching chains, is as good an example as any of how the inventor was meant to be. When Sebastian de Ferranti designed the Deptford power station to provide a large segment of London with electricity, he was portrayed as the modern Colossus, straddling the Thames, a dynamo in one hand and a light bulb in the other. Inventors worked very hard at their image and copy-hungry newspaper men were happy to help. Thomas Edison was proud of his reputation as a tough, hard-working and plain-spoken business inventor. Nikola Tesla assiduously cultivated his image as the other-worldly and iconoclastic eccentric. The Victorian public had been taught to see men of science and invention in some very specific ways.

It is striking just how individualist this public image was – particularly since the reality behind the charismatic men was increasingly collective. Massive engineering exploits like the Atlantic Cable were the work of multitudes of expert workers, even if only a singular name like Cyrus Field's was attached to them in the end. The notion that scientific discovery was the work of talented men, and invention the product of singular minds, was firmly entrenched in the Victorian view of the world and who made it. It is striking, too, how effortlessly this assumption that innovation always emanated from charismatic individuals was transferred into scientific romance. H.G. Wells' cavorite was invented by that charismatic individual Dr Cavor in his own country house laboratory, not in an industrial workshop. One of George Griffith's flying machine inventors built his invention in a lonely garret. E.E. Kellett's lady automaton was assembled by a scientific loner. Invariably, the future explored in Victorian scientific romances was a future made out of the inventions of singular men, just as the future on show at world fairs was made by them, too. So, in fact and fiction, Victorian futures were owned by the charismatic men of invention who had engineered them.

Those imagined futures were often playgrounds for singular men, as well. Dr Cavor's voyage to the Moon was a personal adventure. George Griffith had the hero of his 'Stories of Other Worlds', Lord Redgrave, take a honeymoon in space, jaunting from planet to planet with his bride and his faithful retainer acting as engineer and pilot. John Jacob Astor's protagonist might have been president, but his flying to the Moon and his play among the stars was carried out in an entirely private capacity. Tales like this simply underlined the extent to which the disciplines of empire were part of the Victorian future, too. Astor's President Bearwarden might have been on a safari to the planets, but the Earth he left behind was one divided between globe-spanning English and American empires. Scientific romances featuring heroic individuals exploring the cosmos mirrored stories from the *Boy's Own Paper* where intrepid adventurers trekked through jungles. Lord Redgrave might have been H. Rider Haggard's Sir Henry Curtis – and in an interesting inversion it is worth remembering that Haggard's explorers in *King Solomon's Mines* passed themselves off to the natives as 'white men from the stars'. Explorers in space might just as easily have been the Royal Geographical Society's adventurers, searching for the source of the Nile.

Stories like these, whether set in the interior of Africa, or on the Moon, served to emphasise the disciplines of empire. But the fantasies of imperialist exploits extrapolated to the stars also emphasised the degree to which the Victorian future would be an imperial one, too. International exhibitions made that clear as well. In the aftermath of the Great Exhibition, William Whewell's depiction of it as a frozen tableau of the rise of civilisation made this connection perfectly clear. The spoils of empire were there on show cheek by jowl with the latest technological innovations. In subsequent exhibitions, the resources of Europe's empires were displayed alongside their engineering accomplishments – and those resources included their peoples, too.

International exhibitions routinely featured native villages where visitors could wonder around and experience colonial life at second hand. The organisation of these exhibitions, just like the plots of scientific romances, simply assumed that the logic of European empire would extend into the future. There was a similar logic at play in the illustrations of life 100 years into the future that were commonplace as the century turned. They depicted the fantastic technologies of the year 2000 being used by people dressed in *fin de siècle* fashion. It was another reminder of just whose futures these were.

The science that the Victorians reinvented to get them to that future was a thoroughly imperial science. During his long tenure at the Royal Society, Joseph Banks had presided over a learned empire that mapped quite well onto the physical one. His networks linked the Royal Society to Kew Gardens, the Admiralty, the East India Company and Greenwich Observatory. It was partly to gain control of these extensive networks of imperial power, that Banks had spent his career weaving together, that the new generation of disciplined men aimed to reform the Royal Society. The reformers were themselves men well embedded in commercial and imperial networks. Some of their biggest projects, like the magnetic crusades to map the variations in the Earth's magnetic field, were expressly designed with commerce and empire in mind – to ease the to and fro of trade and power by making maritime navigation easier. The telegraph and the scientific and engineering expertise that underpinned its expansion was – as Lord Salisbury reminded the assembled gathering of electrical engineers in 1889 – one of the most powerful tools of imperial control. Expeditions to observe eclipses and the transit of Venus depended on imperial reach as well as being themselves expressions of what the Empire could do.

All the technological innovations – actual and imagined – described in this book had imperial entanglements. They all depended on the resources that only an imperial power could provide, and they all offered ways of consolidating and expanding the reach of empire. Steam engines powered ships and locomotives, and in the Victorian imagination so would electricity in the future. The electric telegraph, the telephone and wireless telegraphy provided new ways of keeping the Empire together – giving the centre of imperial power tools that would prove vital for keeping the peripheries under control. The never-realised telectroscope offered to provide an all-seeing eye for the state. Babbage's Difference Engine and its Analytical Engine successor, neither ever built, were imagined as ways of dealing mechanically with a growing deluge of numbers, actuarial as well as astronomical. The mechanical calculators that really were developed were essential tools for growing state and imperial bureaucracies. The dream of flying, from George Cayley's experiments onwards, was in part, at least, a dream of empire. The presumption in scientific romances that the domination of the skies was ultimately a means of terrestrial domination is telling, as was reflected in the fear expressed in the aftermath of Louis Blériot's cross-channel flight that Britain was no longer an invincible island.

The ways we think about our future now is the product of this Victorian mix of fact and fantasy, and that is one reason, at least, why it matters that we try and make sense of it. The ways in which the Victorians understood science and its place in their culture, and the ways in which they understood the future, were intimately linked. Science, they understood, was the province of disciplined minds, and particular kinds of disciplined minds at that. Men of science of the right sort could be trusted with nature because they exhibited the right kind of qualities for the job. Increasingly, they were the products of rigorous regimes of training. Cavendish men at Cambridge, and students at other universities that emulated the approach, were inculcated in the discipline of experiment. Accuracy and precision were not just attributes of the measurements they took, they were meant to be moral attributes of the men who took the measurements, as well. They were exemplary individuals. Their science, and their inventions, were the results of their singularity. The Victorians had very clear ideas, by the end of the nineteenth century, of what men of science and their institutions should look like, and what they were for. It was a view that we have inherited.

It was men like these who were going to make the future. And by and large, we still imagine that our future will be made by men like these – charismatic innovators willing to take risks but possessed of the self-discipline and drive needed to get the job done. Just, perhaps, as the Victorians took steam technology for granted and fantasised about electrical futures instead, we place more value on innovation than reliability. We still celebrate Victorian makers of futures, and our inventor–entrepreneurs (who would probably react with horror to being called Victorian) do their best to emulate them – or more accurately, to buy into the myths with which we have surrounded them. We even take sides – was it Edison or Tesla who really invented our future for us? Who are the heroes and who are the villains in the story of how the modern world was made? For their supporters, their ruthless drive towards the future is the contemporary inventor–entrepreneur's key virtue. Their showmanship, just like Edison's or Tesla's showmanship, is a way of demonstrating their iconoclasm. For their detractors, of course, the very same drive is their key vice, but the assumption that these mould-breaking individuals really are the men who own the future runs through their stories too.

The Victorians changed the world by reinventing science. They turned it into a tool for making the future – and they were astonishingly successful in doing so. Thanks to their efforts, the world in 1900 had changed immeasurably from the world of 1800. Someone born in the 1820s would have seen their world change during their lifetime in unimaginable ways – and certainly on a scale that bore no comparison to the past. The Victorians may not have made it to the Moon, but they had the Moon in their sights and they were confident that they would get there. In the meantime, their steam locomotives thundered across land and their dreadnoughts dominated the seas. Within a decade of Queen Victoria's death, their flying machines would be well on the way to dominating the skies, as well. City streets glittered like glowing streams of brilliance through the night. Information flowed ceaselessly through cables and – by the end of the century – through the ether. Theirs really was a world utterly transformed. As they pushed their culture relentlessly into the future, they devoured stories about what that future world would be like. It was a future that would depend, increasingly, on expert scientific knowledge to sustain its technologies.

The world we have inherited from the Victorians is not quite the future they imagined, but it is a world that depends entirely on the hidden work of experts. It is their expertise, not the exploits of iconoclastic inventors of the kind mythologised by the Victorians and still venerated by us, that will get us to the future – if we get there at all. The Victorians saw the future as being the achievement of singular men, rather than hidden armies of expert workers, and thanks to them, we still think this way about our own futures. That matters. It matters because it constrains our visions of how the future might be remade, and who owns that future. By following the Victorian recipe for future-making, without reflecting on the fact that the ways we think are constrained by their histories, we are in danger of limiting access to the future to only particular kinds of people, just as the Victorians did – they thought it only belonged to them. We can't get away from the fact that the science that governs – and may well save – our lives, is the product of an imperial culture. But if we want to change science and its culture for the better, we need to start by remembering this history. After all, if the Victorians transformed science and what it was for, so can we.

Notes

Prologue: Inventing the Future

1 Frederick Bramwell, *Our Big Guns: An Address Delivered in the Town Hall, Birmingham* (London: William Coles and Sons, 1886).

2 Brandy Schillace, *Clockwork Futures: The Science of Steampunk and the Reinvention of the Modern World* (New York: Pegasus Books, 2017).

3 Will Tattersdill, *Science, Fiction, and the Fin-de-Siècle Periodical Press* (Cambridge: Cambridge University Press, 2016).

4 Samuel Smiles, *Self-Help: With Illustrations of Character and Conduct* (London: John Murray, 1859).

5 Peter Bowler, *Progress Unchained: Ideas of Evolution, Human History and the Future* (Cambridge: Cambridge University Press, 2021).

6 John Stuart Mill, 'The Spirit of the Age', *Examiner*, 9 January 1831, p. 20.

7 Simon Schaffer, 'The Nebular Hypothesis and the Science of Progress', in Jim Moore (ed.), *History, Humanity and Evolution* (Cambridge: Cambridge University Press, 1989), pp. 131–64.

8 Adrian Desmond, *The Politics of Evolution: Morphology, Medicine, and Reform in Radical London* (Chicago: University of Chicago Press, 1990).

9 Crosbie Smith, *The Science of Energy: A Cultural History of Energy Physics in Victorian Britain* (London: Athlone Press, 1998).

10 Daniel Pick, *Faces of Degeneration: A European Disorder* (Cambridge: Cambridge University Press, 1989).

11 Sydney Ross, 'Scientist: The Story of a Word', *Annals of Science*, 1962, 19: 65–85.

12 Smiles, *Self-Help*, p. 1.

13 Ibid., p. 2.

14 Letter from the Duke of Somerset to Charles Babbage, quoted in Iwan Rhys Morus, *Michael Faraday and the Electrical Century* (London: Icon Books, 2004), p. 102.

15 Melanie Keene, *Science in Wonderland: The Scientific Fairy Tales of Victorian Britain* (Oxford: Oxford University Press, 2015).

16 Smiles, *Self-Help*, p. 12.

17 John Tyndall, *Faraday as a Discoverer* (London: Longmans, Green & Co., 1868), p. 45.

18 Smiles, *Self-Help*, p. 23.

19 William Robert Grove, *A Lecture on the Progress of Physical Science Since the Opening of the London Institution* (London: London Institution, 1842), p. 37.

20 Alfred Smee, *Elements of Electro-metallurgy* (London: Longman, Rees, Orme, Brown and Longman, 1841), p. 147.

Chapter 1: Science Wars

1 'Death of Sir Joseph Banks', *Morning Post*, 20 June 1820.

2 'President of the Royal Society', *Morning Post*, 24 June 1820.

3 Charles Lyte, *Sir Joseph Banks: 18th Century Explorer, Botanist and Entrepreneur* (London: David & Charles, 1980); Patrick O'Brian, *Joseph Banks: A Life* (London: Harvill Press, 1987); John Gascoigne, *Joseph Banks and the English Enlightenment: Useful Knowledge and Polite Culture* (Cambridge: Cambridge University Press, 2008).

4 Quoted in Lyte, *Sir Joseph Banks*, p. 43.

5 Steven Shapin, *A Social History of Truth: Civility and Science in Seventeenth-Century England* (Chicago: University of Chicago Press, 1994).

6 David Philip Miller, 'Joseph Banks, Empire, and Centres of Calculation in late Hanoverian London', in David Philip Miller and Peter Hanns Reill (eds), *Visions of Empire: Voyages, Botany, and Representations of Nature* (Cambridge: Cambridge University Press, 1996), pp. 21–36.

7 Anne Salmond, *Bligh: William Bligh in the South Seas* (Berkeley: University of California Press, 2011).

8   John Keay, *The Honourable Company: A History of the English East India Company* (London: HarperCollins, 1993); William Dalrymple, *The Anarchy: The Relentless Rise of the East India Company* (London: Bloomsbury, 2019).

9   Gascoigne, *Joseph Banks and the English Enlightenment*, pp. 220–21.

10 'Preface, Advertisement, Address, and a Rare Whack at the Voracious Bats', *The Lancet*, 1831–2, 17: 1–16, p. 2.

11 Cherry Lewis and Simon J. Knell (eds), *The Making of the Geological Society of London* (London: Geological Society, 2009).

12 Horace Woodward, *The History of the Geological Society of London* (London: Longmans, 1908).

13 Martin Rudwick, 'The Foundation of the Geological Society of London: Its Scheme for Cooperative Research and its Struggle for Independence', *The British Journal for the History of Science*, 1963, 1: 325–55.

14 Quoted in ibid., p. 333.

15 Quoted in ibid., p. 344.

16 Quoted in ibid., p. 346.

17 Quoted in Gascoigne, *Joseph Banks and the English Enlightenment*, p. 256.

18 William Ashworth, 'The Calculating Eye: Baily, Herschel, Babbage and the Business of Astronomy', *The British Journal for the History of Science*, 1994, 27: 409–41.

19 Francis Baily to Charles Babbage, 11 March 1820, quoted in Ashworth, ibid., p. 412.

20 John Herschel, *Memoir of Francis Baily, Esq.* (London: Moyes & Barclay, 1845), p. 10.

21 Francis Baily, *Memoir on a New and Certain Method of Ascertaining the Figure of the Earth by Means of Occultations of the Fixed Stars, by A. Cagnoli* (London: Richard and Arthur Taylor, 1819), p. 29.

22 Ibid., pp. 30–31.

23 Ibid., pp. 33–4.

24 Herschel, *Memoir of Francis Baily*, p. 16.

25 Doron Swade, *The Cogwheel Brain: Charles Babbage and the Quest to Build the First Computer* (London: Abacus, 2001).

26 'A Review of Some Leading Points in the Official Character and Proceedings of the Late President of the Royal Society', *Philosophical Magazine*, 1820, 56: 161–74, 241–57, p. 254.

27 Ibid. p. 257.

28 David Miller, 'Between Hostile Camps: Sir Humphry Davy's Presidency of the Royal Society of London, 1820–1827', *The British Journal for the History of Science*, 1983, 16: 1–47.

29 Jan Golinski, *The Experimental Self: Humphry Davy and the Making of a Man of Science* (Chicago: University of Chicago Press, 2016).

30 Quoted in Miller, 'Between Hostile Camps', p. 29.

31 Francis Baily, *Remarks on the Present Defective State of the Nautical Almanac* (London: John Richardson, 1822), p. 18.

32 Ibid., p. 49.

33 Council Minutes of the Royal Society, 3 May 1827, Royal Society Library.

34 Charles Babbage, *Reflections on the Decline of Science in England, and on Some of its Causes* (London: B. Fellowes, 1830), p. 160.

35 *The Times*, 25 November 1830.

36 *The Times*, 25 November 1830.

37 *The Times*, 29 November 1830.

38 *The Times*, 1 December 1830.

39 'News Raisonnee', *The Age*, 5 December 1830, p. 386.

40 *The Times*, 1 December 1830.

41 *Morning Post*, 9 December 1830.

42 Babbage meant the term 'profession' here in the sense of the traditional professions of law, medicine, the Church and the military. Babbage, *Reflections on the Decline of Science in England*.

43 Ibid., p. 11.

44 Ibid., p. 213.

45 Jack Morrell and Arnold Thackray, *Gentlemen of Science: Early Years of the British Association for the Advancement of Science* (Oxford: Oxford University Press, 1982).

46 Frank A.J.L. James, 'Michael Faraday's Work on Optical Glass', *Physics Education*, 1991, 26: 296–300.

47 *Nautical Almanac and Astronomical Ephemeris for the Year 1834* (London: John Murray, 1833).

48 Robert Smith, 'A National Observatory Transformed: Greenwich in the Nineteenth Century', *Journal for the History of Astronomy*, 1991, 22: 5–20.

49 Charles Babbage, *On the Economy of Machinery and Manufactures* (London: John Murray, 1832).

50 George Bellas Greenough, 'Address of the President', *Proceedings of the Geological Society of London*, 1838, 2: 42–70, p. 51.

51 'Report from Messrs. Faraday and Lyell to the Rt. Hon. Sir James Graham, Bart., Secretary of State for the Home Department, on the Subject of the Explosion at the Haswell Collieries, and the Means of Preventing Similar Accidents', *Philosophical Magazine*, 1845, 26: 16–35.

52 James Secord, *Visions of Science: Books and Readers at the Dawn of the Victorian Age* (Chicago: University of Chicago Press, 2014).

53 John Herschel, *Preliminary Discourse on the Study of Natural Philosophy* (London: Longman, Rees, Orme, Brown and Green, 1831), p. 44.

54 Ibid. p. 70; p. 72.

55 Letter from Michael Faraday to William Robert Grove, 21 December 1842, Frank James, *The Correspondence of Michael Faraday* (London: Institution of Engineering and Technology, 1996), vol. 3, p. 113.

56 William Robert Grove, 'New Voltaic Battery Gaseous Elements', *Literary Gazette*, 1842, 26: 833.

57 Morus, *William Robert Grove: Victorian Gentleman of Science* (Cardiff: University of Wales Press, 2017).

58 William Robert Grove, 'Physical Science in England', *Blackwood's Magazine*, 1843, 54: 514–25, p. 517.

59 Council Minutes of the Royal Society, 5 November 1846, Royal Society Library.

60 Spencer Compton, 'Presidential Address', *Proceedings of the Royal Society*, 1847, 8: 698–703, p. 699.

61 Letter from Edward Forbes to William Robert Grove, undated, Royal Institution Library.

Chapter 2: Practical Men

1 'Dinner at the Thames Tunnel', *The Standard*, 12 November 1827.

2 Ralph Dodd, *Reports, with Plans, Sections, &c. of the Proposed Dry Tunnel, or Passage, from Gravesend, in Kent, to Tilbury, in Essex* (London: J. Taylor, 1798), p. 2.

3 David Lampe, *The Tunnel: The Story of the World's First Tunnel Under a Navigable River Dug Beneath the Thames, 1824–42* (London: George G. Harrap & Co., 1963).

4 Carolyn Cooper, 'The Portsmouth System of Manufacture', *Technology and Culture*, 1984, 25: 182–225.

5 'Tunnel under the Thames', *The Mirror of Literature, Amusement, and Instruction*, 22 May 1824, 321–3, p. 322.

6 Richard Beamish, *Memoir of the Life of Sir Marc Isambard Brunel* (London: Longman, Green, Longman and Roberts, 1862), p. 209.

7 'The Thames Tunnel', *The Times*, 19 May 1827.

8 'Accident at the Thames Tunnel', *Morning Chronicle*, 19 May 1827; 'The Thames Tunnel – Most Alarming Accident', *Morning Post*, 19 May 1827.

9 'Late Accident at the Thames Tunnel', *Morning Chronicle*, 21 May 1827.

10 Ibid.

11 'Accident at the Thames Tunnel', *The Times*, 14 January 1828.

12 Quoted in Isambard Brunel, *The Life of Isambard Kingdom Brunel, Civil Engineer* (London: Longman, Green & Co., 1770), p. 27; p. 29.

13 'Opening of the Thames Tunnel', *Morning Chronicle*, 27 March 1843.

14 'To the Mechanics of the British Empire', *Mechanics' Magazine*, 1823, 1: unpaginated.

15 'Memoir of James Watt', *Mechanics' Magazine*, 1823, 1: 1–6, p. 1.

16 'Preface', *Mechanics' Magazine*, 1823, 1: iii–iv, p. iii.

17 David Stack, *Nature and Artifice: The Life and Thought of Thomas Hodgskin, 1787–1869* (London: Royal Historical Society, 1998).

18 Thomas Hodgskin, *Labour Defended Against the Claims of Capital* (London: Knight & Lacey, 1825).

19 David J. Jeremy, 'Damming the Flood: British Government Efforts to Check the Outflow of Technicians and Machinery, 1780–1843', *Business History Review*, 1977, 51: 1–34.

20 Andrew Ure, *The Philosophy of Manufactures: Or, an Exposition of the Scientific, Moral, and Commercial Economy of the Factory System of Great Britain* (London: Charles Knight, 1835), p. 19.

21 'Institutions for Instruction of Mechanics', *Mechanics' Magazine*, 1823, 1: 99–102, pp. 99–100.

22 'London Mechanics' Institution. Fourth Quarterly Meeting', *Mechanics'*
    Magazine, 1824, 3: 187–92, p. 191.

23 'Public Meeting for the Establishment of the London Mechanics' Institute',
    Mechanics' Magazine, 1823, 1: 177–92, p. 190.

24 'London Mechanics' Institution', *Mechanics' Magazine*, 1823, 1: 227–29, p. 229.

25 'London Mechanics' Institution', *Mechanics' Magazine*, 1824, 2: v–viii, p. vii,
    p. viii.

26 'Preface', *Mechanics' Magazine*, 1829, 9: iii–vi, p. vi.

27 Isambard Brunel, *The Life of Isambard Kingdom Brunel*, pp. 1–2.

28 Smiles, *Self-Help*, p. 5.

29 Ibid., p. 26.

30 Boyd Hilton, *The Age of Atonement: The Influence of Evangelicalism on Social*
    and Economic Thought, 1785–1864 (Oxford: Oxford University Press, 1988).

31 Quoted in Samuel Smiles, *The Life of Thomas Telford, Civil Engineer* (London:
    John Murray, 1867), p. 134.

32 Ibid., p. 162.

33 Thomas Roscoe, *The London and Birmingham Railway* (London: Charles
    Tilt, 1839), unpaginated prefatory notice.

34 Ibid., p. 3.

35 Quoted in Brunel, *The Life of Isambard Kingdom Brunel*, p. 82.

36 Ibid., p. 233.

37 Ibid., p. 314.

38 Samuel Smiles, *The Life of George Stephenson, Railway Engineer*, 3rd ed.
    (London: John Murray, 1857), p. 116.

39 Ibid., p. 117.

40 John Ayrton Paris, *The Life of Sir Humphry Davy, Bart.* (London: Henry
    Colburn & Richard Bentley, 1831), vol. 2, p. 107.

41 Ibid., p. 129.

42 Morus, *Michael Faraday and the Electrical Century*, pp. 52–9.

43 Poster, illustrated in Michael Hoskin, 'Astronomers at War: South v.
    Sheepshanks', *Journal of the History of Astronomy*, 1989, 20: 175–212, p. 192.

44 Ibid., p. 198.

45 Morus, *Frankenstein's Children: Electricity, Exhibition and Experiment in Early Nineteenth-Century London* (Princeton: Princeton University Press, 1998), pp. 198–220.

46 William F. Cooke, *The Electric Telegraph: Was it Invented by Professor Wheatstone?* (London: W.H. Smith and Son, 1856–7), vol. 1, pp. 16–17.

47 'Mr Bain's Electro Magnetic Inventions', *Mechanics' Magazine*, 1843, 39: 64–77.

48 'Books on Electrometallurgy', *Mechanics' Magazine*, 1842, 36: 458–62, p. 458; George Shaw, *A Manual of Electro-metallurgy* (London: Simpkin, Marshall, and Co., 1842); Alfred Smee, *Elements of Electro-metallurgy, or the Art of Working in Metals by the Galvanic Fluid* 2nd ed. (London: E. Palmer, 1842).

49 'Books on Electrometallurgy', p. 458, p. 459.

50 William Robert Grove, 'Physical Science in England', *Blackwood's Magazine*, 1843, 23: 514–25, p. 521.

Chapter 3: Measure for Measure

1 Robert Monro Black, *The History of Electric Wires and Cables* (London: Peter Peregrinus, 1983), p. 26.

2 Quoted in Isabella Field Judson, *Cyrus W. Field: His Life and Work* (New York: Harper & Brothers, 1896), p. 90.

3 Ibid., p. 41.

4 Andrea Wulf, *The Invention of Nature: The Adventures of Alexander von Humboldt, the Lost Hero of Science* (London: John Murray, 2015).

5 Bruce Hunt, 'Scientists, Engineers, and Wildman Whitehouse: Measurement and Credibility in Early Cable Telegraphy', *The British Journal for the History of Science*, 1996, 29: 155–69.

6 Edward Wildman Whitehouse, *Experimental Observations on Submarine Electric Cables* (Brighton: Arthur Wallis, 1855), p. 5.

7 Judson, *Cyrus W. Field*, p. 95.

8 Ibid., p. 96.

9 Ibid., p. 101.

10 Ibid., p. 112.

11 'The Atlantic Cable', *The Times*, 6 September 1858, p. 7.

12 Crosbie Smith and Norton Wise, *Energy and Empire: A Biographical Study of Lord Kelvin* (Cambridge: Cambridge University Press, 1989).

13 Crosbie Smith, *The Science of Energy: A Cultural History of Energy Physics in Victorian Britain* (London: Athlone Press, 1998).

14 Judson, *Cyrus W. Field*, p. 154.

15 Ibid., p. 207.

16 'The Atlantic Cable', *The Times*, 28 July 1866, p. 9.

17 Michael Barton, Geoffrey Belknap, Iwan Rhys Morus and James Ungureanu (eds), *The Correspondence of John Tyndall* (Pittsburgh: Pittsburgh University Press, 2022), vol. 9, p. 116.

18 William Robert Grove, 'Presidential Address', *Report of the British Association for the Advancement of Science*, 1866, 26: liii–lxxxi, p. lxvi.

19 Smith and Wise, *Energy and Empire*, pp. 120–34.

20 Robert Kalley Miller, 'The Proposed Chair of Natural Philosophy', *Cambridge University Reporter*, 23 November 1870, pp. 118–19, quoted in Simon Schaffer, 'Late Victorian Metrology and its Instrumentation: A Manufactory of Ohms', Robert Bud and Susan Cozzens (eds), *Invisible Connections: Instruments, Institutions, and Science* (Bellingham: SPIE Optical Engineering Press, 1992), 23–56, p. 24.

21 William Thomson, 'Electrical Units of Measurement', in *Popular Lectures and Addresses* (London: Macmillan, 1888–94), vol. 1, pp. 80–143.

22 Latimer Clark and Charles Bright, 'On the Formation of Standards of Electrical Quantity and Resistance', *Report of the British Association for the Advancement of Science*, 1862, 31: p. 37.

23 Smith, *The Science of Energy*, pp. 211–38; Lewis Campbell and William Garnett, *The Life of James Clerk Maxwell* (London: Macmillan, 1882).

24 James Clerk Maxwell and Fleeming Jenkin, 'On the Elementary Relations between Electrical Measurements', *Report of the British Association for the Advancement of Science*, 1863, 32: 130–63, p. 131.

25 James Clerk Maxwell to Edward Blore 15 February 1871, in Campbell and Garnett, *The Life of James Clerk Maxwell*, pp. 265–6.

26 Ibid., p. 244.

27 James Clerk Maxwell, 'Introductory Lecture on Experimental Physics' in W.D. Niven (ed.), *The Scientific Papers of James Clerk Maxwell* (Cambridge: Cambridge University Press, 1890), vol. 2, 241–55, p. 242.

28 James Clerk Maxwell, 'Electricity and Magnetism', *Nature*, 15 May 1873, 42–3, p. 42.

29 Robert Strutt, *The Life of Lord Rayleigh* (London: Edward Arnold & Co., 1924).

30 Simon Schaffer, 'Physics Laboratories and the Victorian Country House', in Crosbie Smith and Jon Agar (eds), *Making Space for Science: Territorial Themes in the Shaping of Knowledge* (Oxford: Oxford University Press, 1982), pp. 22–4.

31 Strutt, *The Life of Lord Rayleigh*, p. 105.

32 Quoted in ibid., p. 109.

33 J.J. Thomson, *Recollections and Reflections* (London: Macmillan, 1936), p. 95, p. 123.

34 Quoted in David Cahan, 'The Institutional Revolution in German Physics, 1865–1914', *Historical Studies in the Physical Sciences*, 1985, 15: 1–65, p. 23.

35 Quoted in David Cahan, 'Werner Siemens and the Origins of the Physikalisch-Technische Reichsanstalt', *Historical Studies in the Physical Sciences*, 1982, 12: 253–83, p. 254.

36 Quoted in Robert Fox, *The Savant and the State: Science and Cultural Politics in Nineteenth-Century France* (Baltimore: Johns Hopkins University Press, 2012), p. 255.

37 Charles Bristed, *Five Years in an English University* (New York: 1852), quoted in Andrew Warwick, 'Exercising the Student Body: Mathematics and Athleticism in Victorian Cambridge', Christopher Lawrence and Steven Shapin (eds), *Science Incarnate* (Chicago: University of Chicago Press, 1998), p. 295.

38 Quoted in George Sweetnam, 'Precision Implemented: Henry Rowland, the Concave Diffraction Grating, and the Analysis of Light', M. Norton Wise (ed.), *The Values of Precision* (Princeton: Princeton University Press, 1995), p. 284.

39 Quoted in Strutt, *The Life of Lord Rayleigh*, p. 117.

40 *First Report of the Royal Commissioners on Technical Instruction* (London: Her Majesty's Stationery Office, 1882), p. 3.

41 Smith and Wise, *Energy and Empire*.

42 Gerald Geison, *Michael Foster and the Cambridge School of Physiology: The Scientific Enterprise in Late Victorian Society* (Princeton: Princeton University Press, 1978).

43 Jules Verne, *Around the World in Eighty Days*, originally published in 1873 (Harmondsworth: Penguin, 2008), p. 3.

44 Garnet J. Wolseley, *The Soldier's Pocket-Book for Field Service*, 3rd ed. (London: William Clowes and Sons, 1874), pp. 13–14.

45 Arthur Conan Doyle, 'A Study in Scarlet', originally published 1887, in *The Adventures of Sherlock Holmes* (Ware: Wordsworth Classics, 1992), p. 12.

46 John Cawood, 'The Magnetic Crusade: Science and Politics in Early Victorian Britain', *Isis*, 1979, 70: 492–518.

47 William Dalrymple, *The Anarchy* (London: Bloomsbury, 2019).

48 Rudyard Kipling, *Kim*, originally published in 1901 (Ware: Wordsworth Classics, 1993), p. 146.

49 'The Eclipse Expedition in India', *Illustrated London News*, 10 January 1872.

50 'The Eclipse Expedition in India', *Illustrated London News*, 27 January 1872.

51 George Biddell Airy, *Account of Observations of the Transit of Venus* (London: Her Majesty's Stationery Office, 1881), p. 9.

52 John Beddoe, *The Races of Great Britain* (London: Trubner & Co., 1862).

53 Chandak Sengoopta, *Imprint of the Raj: How Fingerprinting was Born in Colonial India* (London: Macmillan, 2003).

54 Pierre Duhem, *The Aim and Structure of Physical Theory* (Princeton: Princeton University Press, 1954), pp. 70–71.

Chapter 4: Showing Off

1 *National History and Views of London and its Environs* (London: Allam Fell & Co., 1834), p. 30.

2 *Catalogue, National Gallery of Practical Science, Blending Instruction with Amusement* (London: J. Holmes, 1832), p. 1.

3 'Exhibition of Works of Popular Science', *Mechanics' Magazine*, 1832, 7: 158–60, p. 158.

4 'Mr Perkins' Extraordinary Steam Gun', *The London Mechanics' Register*, 1824, 1: 1–4, p. 3.

5 Ibid., p. 4.

6 'Steam', *New Monthly Magazine*, 1825, 13: 194–5, p. 194.

7 Samuel Goodrich, *Recollections of a Lifetime*, two volumes (New York: Miller, Orton and Mulligan, 1826), vol. 2, p. 129.

8 Charles Coleman Sellers, *Mr Peale's Museum: Charles Willson Peale and the First Popular Museum of Natural Science and Art* (New York: Norton, 1980).

9 Charles Babbage, *Passages from the Life of a Philosopher* (London: Longman, Green and Co., 1864).

10 'Specification of the Patent Awarded to Mr Robert Barker, of Edinburgh', *The Repertory of Arts and Manufactures*, 1796, 4: 165–67, p. 166.

11 Ibid, p. 167.

12 'Lyceum, Strand', *Morning Chronicle*, 1 October 1801.

13 'Advertisement to the Public', *Transactions of the Society Instituted at London for the Encouragement of Arts, Manufactures, and Commerce*, 1828, 46: unpaginated.

14 'Public Exhibition of British Manufacture', *Mechanics' Magazine*, 1828, 9: 195–6.

15 Ibid., p. 196.

16 'National Repository of Arts and Sciences, King's Mews', *Morning Post*, 24 June 1828, p. 3.

17 'The National Repository', *The Times*, 21 July 1828, p. 2.

18 'National Repository, Charing Cross', *Literary Gazette*, 1828, 12: 410.

19 'National Repository', *Mechanics' Magazine*, 1829, 11: 58.

20 'The National Repository', *The Times*, 2 March 1829.

21 'A Revival and a Removal: The National Repository', *Mechanics' Magazine*, 1833, 19: 336.

22 'Henry's European Diary', 20 March 1837, in Nathan Reingold et al. (ed.), *The Papers of Joseph Henry* (Washington DC: Smithsonian Institution Press, 1972), vol. 3, p. 179. 'March' refers to the instrument maker James Marsh.

23 Smithsonian Institution Archives, Record Unit 7056, Joseph Saxton Papers.

24 'Gallery of Practical Science', *Literary Gazette*, 1835, 19: 315.

25 Albert Smith, *Gavarni in London: Sketches of Life and Character* (London: David Bogue, 1849), p. 13.

26 Quoted in Toshio Kusamitsu, 'Great Exhibitions before 1851', *History Workshop*, 1980, 9: 70–89, p. 80.

27 *Catalogue of the National Gallery of Practical Science, Blending Instruction with Amusement* (London: Society for the Encouragement of Practical Science, 1832), p. 34, p. 43.

28 Quoted in Brenda Weeden, *The Education of the Eye: History of the Royal Polytechnic Institution, 1838–81* (London: Granta Editions, 2008), p. 12.

29 Jehangeer Nowrojee and Hirjeebhoy Merwanjee, *Journal of a Residence of Two Years and a Half in Great Britain* (London: W.H. Allen, 1841), p. 120.

30   Ibid., p. x.

31   'The Royal Polytechnic Institution', *Morning Chronicle*, 16 September 1843.

32   Quoted in J.J. Fahie, *A History of the Electric Telegraph to the Year 1837* (London: E. & F.N. Spon, 1884), p. 418.

33   'The Royal Panopticon of Science and Art', *Morning Post*, 17 March 1854, p. 5.

34   Quoted in Robert Kargon, *Science in Victorian Manchester: Enterprise and Expertise* (Baltimore: Johns Hopkins University Press, 1977), p 37

35   'Opening of the Royal Victoria Gallery', *The Manchester Times*, 13 June 1840, p. 2.

36   Benjamin Love, *The Hand-book of Manchester* (Manchester: Love and Barton, 1842), p. 218.

37   'The Great Induction Coil at the Polytechnic Institution', *The Times*, 7 April 1869, p. 4.

38   'Playing with Lightning', *All the Year Round*, 1869, 1: 617–620, p. 620.

39   'The Industrial Exhibition', *The Child's Companion: or Sunday Scholar's Reward*, 1 April 1851, 100–104, p. 100.

40   'The Crystal Palace', *Lady's Newspaper and Pictorial Times*, 18 January 1851, 1–2, p. 1.

41   Editorial, *Daily News*, 2 May 1851, p. 1.

42   'The Great Exhibition', *Daily News*, 22 May 1851, p. 5.

43   Henry Mayhew, *1851: or, The Adventures of Mr and Mrs. Sandboys and Family* (London: David Bogue, 1851), p. 160.

44   William Dalrymple and Anita Anand, *Koh-i-Noor: A History of the World's Most Infamous Diamond* (London: Bloomsbury, 2018).

45   Clement Shorter, *The Brontës: Life and Letters* (London: Hodder & Stoughton, 1908), vol. 2, p. 216.

46   William Whewell, 'On the General Bearing of the Great Exhibition', *Lectures on the Results of the Great Exhibition* (London: David Bogue, 1852), p. 11.

47   George Clayton, *Sermons on the Great Exhibition* (London: Benjamin L. Green, 1851), p. 27.

48   Editorial, *Electrical Review*, 1892.

49   'The Opening of the International Exhibition', *Jackson's Oxford Journal*, 3 May 1862, p. 8.

50 'Opening of the Vienna Exhibition', *Morning Post*, 2 May 1873, p. 5.

51 Charles Gindriez and James M. Hart, *International Exhibitions: Paris–Philadelphia–Vienna* (New York: A.S. Barnes & Co., 1878), p. 22.

52 'Opening of the Melbourne Exhibition', *Daily News*, 2 October 1880, p. 5.

53 'The Paris Electrical Exhibition', *The Standard*, 4 August 1881, p. 5.

54 Quoted in Paul Israel, *Edison: A Life of Invention* (New York: John Wiley & Sons, 1998), p. 214.

55 'Electricity at the Crystal Palace', *Morning Post*, 4 January 1882, p. 3.

56 Editorial, *The Standard*, 8 January 1892, p. 5.

57 'The Eiffel Tower', *Morning Post*, 1 April 1889, p. 5.

58 'The Paris Exhibition', *Morning Post*, 20 April 1889, p. 5.

59 'The Paris Exhibition', *Morning Post*, 7 May 1889, p. 5.

60 Trumbull White and William Igleheart, *The World's Columbian Exposition* (Boston: John K. Hasting, 1893), p. 322.

61 Ibid., p. 301.

62 Candace Wheeler, 'A Dream City', *Harper's Magazine*, 1893, 86: 830–46, p. 846.

63 'At the Fair', *Century Magazine*, 1893, 46: 3–21, p. 7.

Chapter 5: Fuelling the Future

1 'Grand Mechanical Competition – Rail-road Race for 500l', *The Standard*, 9 October 1829.

2 Henry Booth, *An Account of the Liverpool and Manchester Railway* (Liverpool: Wales and Baines, 1831), p. 12.

3 Ibid., p. 75.

4 Ibid.

5 C.H. Greenhow, *An Exposition of the Danger and Deficiencies of the Present Mode of Railway Construction: With Suggestions for its Improvement* (London: John Weale, 1846), p. 421.

6 Dionysius Lardner, 'Inland Transport', *Edinburgh Review*, 1833, 56: 99–146, p. 104.

7 Wolfgang Schivelbusch, *The Railway Journey: The Industrialization and Perception of Time and Space in the Nineteenth Century* (Berkeley and Los Angeles: University of California Press, 1986).

8 Quoted in 'Davenport's Electro-magnetic Machine', *Mechanics' Magazine*, 1837, 27: 404–5.

9   William Robert Grove, *A Lecture on the Progress of Physical Science*, p. 24.

10 Henry's European Diary, 8 April 1837. Nathan Reingold (ed.), *The Papers of Joseph Henry* (Washington DC: Smithsonian Institution Press, 1972), vol. 4, p. 250.

11 William Sturgeon, 'A General Outline of the Various Theories which have been Advanced for the Explanation of Terrestrial Magnetism', *Annals of Electricity*, 1837, 1: 117–23, p. 123.

12 William Sturgeon, 'On Electromagnetism', *Philosophical Magazine*, 1824, 64: 242–9.

13 'Improved Electro-magnetic Apparatus', *Transactions of the Royal Society of Arts*, 1825, 43: 37–52.

14 'National Gallery of Practical Science', *Literary Gazette*, 1833, 17: 730.

15 Edward M. Clarke, 'Description of E.M. Clarke's Magnetic Electric Machine', *Philosophical Magazine*, 1836, 9: 262–6, p. 264.

16 Joseph Saxton, 'Mr J. Saxton on his Magneto-electric Machine; with Remarks on Mr E.M. Clarke's Paper in the Preceding Number', *Philosophical Magazine*, 1836, 9: 360–65, p. 365.

17 Edward M. Clarke, 'Correspondence', *Philosophical Magazine*, 1837, 10: 455.

18 Henry M. Noad, *Lectures on Electricity* (London: George Knight, 1844), p. 380.

19 Clarke, 'Description of E.M. Clarke's Magnetic Electric Machine', p. 264.

20 Noad, *Lectures on Electricity*, p. 381.

21 Walter Rice Davenport, *Thomas Davenport: Pioneer Inventor* (Montpelier: Vermont Historical Society, 1929).

22 Albert Moyer, *Joseph Henry: The Rise of an American Scientist* (Washington DC: Smithsonian Institution Press, 1997).

23 'Electrical Society', *Literary Gazette*, 1838, 22: 458.

24 Jan Golinski, *The Experimental Self: Humphry Davy and the Making of a Man of Science* (Chicago: University of Chicago Press, 2016), p. 111.

25 John Frederic Daniell, 'On Voltaic Combinations', *Philosophical Transactions*, 1836, 126: 107–24.

26 William Robert Grove, 'On a New Voltaic Battery, and on Voltaic Combinations and Arrangements', *Philosophical Magazine*, 1839, 14: 287–93.

27 John Shillibeer, 'Description of a New Arrangement of the Voltaic Battery and Pole Director', *Annals of Electricity*, 1837, 1: 224–5.

28 Moritz Hermann Jacobi to Michael Faraday, 21 June 1839, Frank James (ed.), *The Correspondence of Michael Faraday* (London: Institution of Electrical Engineers, 1993), vol. 2, pp. 590–93.

29 William Robert Grove, 'Friday-Evening Meetings at the Royal Institution', *Philosophical Magazine*, 1840, 26: 338–9.

30 Benjamin Hill, 'On a New Electro-magnetic Machine', *Proceedings of the London Electrical Society*, 1841, 2: 83–6.

31 'British Association – Swansea', Pamphlet, Royal Institution of South Wales.

32 William Robert Grove, 'On the Application of Voltaic Ignition in Lighting Mines', *Philosophical Magazine*, 1845, 27: 442–6.

33 'Electrical Soirée', *Literary Gazette*, 1843, 27: 352.

34 'Electric Light', *Patent Journal*, 1849, 6: 80.

35 'The New Ballet of "Electra" at Her Majesty's Theatre', *Illustrated London News*, 5 May 1849, p. 13.

36 'Staite's Improvements in Lighting', *Patent Journal*, 1848, 4: 169–73.

37 'Taylor's Electro-magnetic Engine', *Mechanics' Magazine*, 1840, 32: 693–6.

38 'The Applicability of Electro-magnetism as a Moving Power', *Inventor's Advocate*, 1820, 2: 410–11.

39 'Taylor's Electro-magnetic Engine'.

40 'Electro-magnetic Power', *The Railway Times*, 1842, 5: 1342.

41 'The Applicability of Electro-magnetism as a Moving Power'.

42 James Prescott Joule, 'Investigations in Magnetism and Electro-magnetism', *Annals of Electricity*, 1849, 4: 131–7, p. 134.

43 James Prescott Joule and William Scoresby, 'Experiments and Observations on the Mechanical Powers of Electro-magnetism, Steam, and Horses', *Philosophical Magazine*, 1846, 28: 448–55.

44 William Robert Grove, 'On the Progress Made in the Application of Electricity as a Motive Power', *Literary Gazette*, 1844, 28: 113.

45 Carlo Matteucci, 'On the Electricity of Flame, with Comments by Mr Grove', *Philosophical Magazine*, 1854, 8: 399–404.

46 William Robert Grove, 'Presidential Address', *Report of the British Association for the Advancement of Science*, 1866, 36: liii–lxxxi.

47 John Tyndall to William Stanley Jevons, 23 May 1866, Barton, Belknap, Morus and Ungureanu (eds), *The Correspondence of John Tyndall*, p. 349.

48 John Tyndall to William Stanley Jevons, 2 June 1866, ibid., p. 359.

49 'The Brayton Ready Motor or Hydrocarbon Engine', *Scientific American*, 1876, 34: 303.

50 'Petroleum Changed into Electricity and Light', *Scientific American*, 1876, 35: 289.

51 'A Petroleum Motor Tricycle', *Scientific American*, 1891, 64: 96.

52 Edward Bulwer-Lytton, *The Coming Race*, originally published in 1871 (London: Gateway, 2015), p. 41.

53 Jules Verne, *Journey to the Centre of the Earth*, originally published in 1864 (Harmondsworth: Penguin, 1994), p. 111.

54 Jules Verne, *Twenty Thousand Leagues under the Sea*, originally published in 1869 (Harmondsworth: Penguin, 1994), p. 86.

55 Albert Robida, *The Twentieth Century*, originally published in 1883 (Middletown CT: Wesleyan University Press, 2004).

56 'The Coming Force', *Punch's Almanack for 1882*, 6 December 1881.

57 John Munro, *The Romance of Electricity* (London: Religious Tract Society, 1893), p. 296, pp. 301–3.

58 John Jacob Astor, *A Journey in Other Worlds: A Romance of the Future* (New York: D. Appleton and Co., 1894), p. 35.

59 Trumbull White and William Igleheart, *The World's Columbian Exposition* (Boston: John K. Hastings, 1893), p. 301.

60 'In 1988: What England will be like a Hundred Years hence', *Answers*, 1 September 1888, p. 15.

61 Henry Wilde, 'Experimental Researches in Electricity and Magnetism', *Philosophical Transactions*, 1867, 157: 89–107, pp. 102–3; p. 106.

62 John Tyndall, 'The Electric Light', *Popular Science Monthly*, 1879, 14: 553–72, p. 560. The Brethren of Trinity House was the committee responsible for overseeing Britain's lighthouses.

63 C.W. Siemens, 'On the Conversion of Dynamical into Electrical Force without the Aid of Permanent Magnetism', *Proceedings of the Royal Society*, 1867, 15: 367–9, p. 368.

64 Charles Wheatstone, 'On the Augmentation of the Power of a Magnet by reaction thereon of Currents induced by the Magnet itself', *Proceedings of the Royal Society*, 1867, 15: 369–372, p. 370.

65 'The Society of Engineers', *The Times*, 2 October 1889, p. 4.

66 'Electric Lighting in London', *The Times*, 3 May 1889, p. 13.

67 White and Igleheart, *The World's Columbian Exposition*, p. 310.

68 Arthur Kennelly, 'Electricity in the Household', *Scribner's Magazine*, 1890, 7: 102–15, p. 115.

69 'Dinner of the Institution of Electrical Engineers', *Electrician*, 1889, 24: 12–15, p. 13.

70 William Pole, *The Life of Sir William Siemens* (London: John Murray, 1888), p. 249.

71 Coleman Sellers, 'The Utilization of Niagara's Power', W.D. Howells, Mark Twain and Nathaniel Shaler, *The Book of Niagara* (Buffalo: Underhill and Nichols, 1893), 193–220, p. 193.

72 George Forbes, 'Harnessing Niagara', *Blackwood's Magazine*, 1895, 158: 430–44, p. 443.

73 Ibid.

74 Sellers, op. cit., note 66, p. 219.

75 Morus, *Nikola Tesla and the Electrical Future* (London: Icon Books, 2019).

76 Thomas Commerford Martin, 'Tesla's Oscillator and Other Inventions', *Century Magazine*, 1895, 9: 916–33, p. 916.

77 'A Hungarian Wizard', *Pall Mall Gazette*, 22 June 1900, p. 4.

78 'What Mr Tesla is Said to Have Said', *Western Electrician*, 14 March 1903, p. 211.

79 'Tesla's Flashes Startling', *New York Sun*, 17 July 1903.

80 William Thomson to James Clerk Maxwell, 24 August 1872, in Peter Harman (ed.), *Scientific Letters and Papers of James Clerk Maxwell* (Cambridge: Cambridge University Press, 1995), vol. 2, p. 749.

81 'Electric Tramway Cars', *The Times*, 6 March 1882, p. 6.

82 'Radium', *The New York Times*, 22 February 1903.

Chapter 6: Surveillance

1 'The Institution of Electrical Engineers', *The Times*, 5 November 1889, p. 7.

2 'Lord Salisbury on Electricity and Civilization', *Daily News*, 5 November 1889.

3 'Dinner of the Institution of Electrical Engineers', *Electrician*, 1889, 24: 12–15, p. 13.

4 Ibid., p. 12.

5 The suggestions were not published at the time, but the document can be found in William Fothergill Cooke, *The Electric Telegraph: Was It Invented by Professor Wheatstone* (London: W.H. Smith and Son, 1856), vol. 2, 239–64, pp. 250–51.

6 Latimer Clark, 'Inaugural Address', *Journal of the Society of Telegraph Engineers*, 1875, 4: 1–22, p. 2.

7 Preface, *Patent Journal and Inventor's Advocate*, 1850, 10: iii–iv, p. iv.

8 Andrew Wynter, 'The Electric Telegraph', *Quarterly Review*, 1854, 95: 118–61, p. 132.

9 'Dinner of the Institution of Electrical Engineers', p. 13.

10 Geoffrey Hubbard, *Cooke and Wheatstone and the Invention of the Electric Telegraph* (London: Routledge, 1965), pp. 105–12.

11 'The Electric Telegraph', *Patent Journal*, 1850, 4: 229–31, p. 231.

12 'Electric Telegraph Company vs. Wilmer and Smith', *The Times*, 26 November 1849, p. 3.

13 'Alleged Misconduct of the Electric Telegraph Company', *Morning Herald*, 11 October 1849, p. 5.

14 F.C. Mather, 'The Railways, the Electric Telegraph and Public Order during the Chartist Period', *History*, 1953, 38: 40–53, p. 50.

15 Ure, *The Philosophy of Manufactures*, p. 19.

16 'Time and the Electric Telegraph', *Mechanics' Magazine*, 1845, 42: 416.

17 'Tricks of the Electrics', *Punch*, 1854, 27: 64.

18 *Proceedings of the Colonial Conference* (London: Her Majesty's Stationery Office, 1887), vol. 1, p. 215.

19 John Tully, 'A Victorian Ecological Disaster: Imperialism, the Telegraph, and Gutta-Percha', *Journal of World History*, 2009, 20: 559–79.

20 'Dinner of the Institution of Electrical Engineers', *Electrician*, p. 12.

21 Vinisha Vinoy, 'The Telephones Patent', https://iopener.today/daily-digest/on-this-day/the-telephones-patent/ [accessed 13 July 2022].

22 'Audible Speech by Telegraph', *Scientific American*, 1876, Supplement vol. 1: 765.

23 'Stentor Distanced', *The Times*, 28 February 1877, p. 6.

24 'The Telephone', *The Times*, 2 July 1877, p. 4.

25 'The British Association', *The Times*, 22 August 1877, p. 11.

26 'Court Circular', *The Times*, 16 January 1878, p. 8.

27 'The Telephone at the Paris Opera', *Scientific American*, 31 December 1881, p. 422.

28 'International Exhibition of Electricity at Paris', *The Times*, 11 August 1881, p. 10.

29 'The Telephonic Central Office System', *Scientific American*, 10 January 1880, p. 15.

30 'The Future of the Telephone', *Scientific American*, 10 January 1880, p. 16.

31 'The Telephone Girl Again', *Electrical Review*, 10 August 1889, p. 6.

32 'The Future of the Telephone', *Scientific American*.

33 Ibid.

34 Electrician, 'The Electroscope', *New York Sun*, 29 March 1877.

35 'The Telectroscope', *The Times*, 27 January 1879.

36 George R. Carey, 'Seeing by Electricity', *Scientific American*, 1880, 44: 355.

37 Willoughby Smith, 'Effect of Light on Selenium During the Passage of an Electric Current', *Nature*, 1873, 7: 303.

38 John Perry and William Ayrton, 'Seeing by Electricity', *Nature*, 1880, 21: 589.

39 Albert Robida, *The Twentieth Century*, trans. Philippe Willems (Middletown CT: Wesleyan University Press, 2004), p. 52.

40 Jan Szczepanik and Ludwig Kleinberg, 'Method and Apparatus for Reproducing Pictures and the Like at a Distance by Means of Electricity', British Patent no. 5031, awarded 24 February 1898.

41 'Vienna', *The New York Times*, 19 March 1898.

42 'An Interesting Invention', *Western Mail*, 24 September 1898.

43 Mark Twain, 'From the "London Times" of 1904', *Century Magazine*, 1898, 57: 100–104, p. 101.

44 Ibid., p. 100.

45 'The Fernseher Again', *Electrical Engineer*, 1898.

46 Oliver Lodge, *Past Years: An Autobiography* (London: Hodder & Stoughton, 1931), pp. 185–6.

47 William Crookes, 'Some Possibilities of Electricity', *Fortnightly Review*, 1892, 51: 173–81, p. 175.

48 E.C. Baker, *Sir William Preece FRS: Victorian Engineer Extraordinary* (London: Hutchinson, 1976), p. 270.

49 J.J. Fahie, *A History of Wireless Telegraphy* (London: Blackwoods, 1899), p. 218.

50 'This Morning's News', *Daily News*, 10 June 1897, p. 1.

51 'Progress of Science', *Graphic*, 12 June 1897, p. 730.

52 'Wireless Telegraphy', *Daily News*, 20 March 1899, p. 5.

53 'A Submarine Destroyer that Really Destroys', *New York Journal*, 13 November 1898.

54 'We May Signal to Mars', *New York Sun*, 25 March 1896.

55 'Tesla's Task of Taming Air', *Chicago Times-Herald*, 15 May 1899.

56 'Has Nikola Tesla Spoken with Mars?', *New York Journal*, 1 January 1901.

57 Louis Pope Gratacap, *The Certainty of a Future Life on Mars* (New York: Irving Press, 1903).

58 'Wireless Telegraphy across the Atlantic', *The Times*, 16 December 1901, p. 5.

59 Ibid., p. 5.

60 William James Copleston, *Memoir of Edward Copleston, D.D., Bishop of Llandaff* (London: John W. Parker and Son, 1851), p. 169.

61 'Dinner of the Institution of Electrical Engineers', *Electrician*, p. 13.

Chapter 7: Calculating People

1 Babbage, *Passages from the Life of a Philosopher*, p. 17.

2 Simon Schaffer, 'Babbage's Dancer and the Impresarios of Mechanism', Francis Spufford and Jenny Uglow (eds), *Cultural Babbage: Technology, Time and Invention* (London: Faber & Faber, 1996), pp. 53–80.

3 Jessica Rifkin, 'The Defecating Duck, or, the Ambiguous Origins of Artificial Life', *Critical Inquiry*, 2003, 29: 599–33.

4 Babbage, *Passages from the Life of a Philosopher*, p. 27.

5 Ibid., p. 36.

6 *Chess: A Selection of Fifty Games, From Those Played by the Automaton Chess-Player, During its Exhibition in London, in 1820* (London: W. People, 1820).

7 'Chess Problems', *The Lancet*, 9 November 1823, 209–10, p. 209.

8 Babbage, *Passages from the Life of a Philosopher*, p. 28.

9 Quoted in Doron Swade, *The Cogwheel Brain: Charles Babbage and the Quest to Build the First Computer* (London: Abacus, 2000), p. 10.

10   Charles Babbage, *A Letter to Sir Humphry Davy on the Application of Machinery to the Purpose of Calculating and Printing Mathematical Tables* (London: J. Booth, 1822), p. 3.

11   Ibid., p. 10.

12   Lorraine Daston, 'Enlightenment Calculations', *Critical Inquiry*, 1994, 21: 182–202.

13   'Royal Institution', *Morning Chronicle*, 29 March 1823.

14   John Herschel, 'Calculating Machinery', *The Times*, 19 August 1828.

15   Charles Babbage, 'Report on the Calculating Engine', 1830, British Library, Add. MSS 37185, fol. 264, quoted in Simon Schaffer, 'Babbage's Intelligence: Calculating Engines and the Factory System', *Critical Inquiry*, 1994, 21: 203–27, p. 218.

16   Ure, *The Philosophy of Manufactures*, p. 14.

17   Dionysius Lardner, 'On Babbage's Calculating Engine', *Edinburgh Review*, 1834, 59: 263–327, p. 264.

18   Ibid., p. 266.

19   Ibid., p. 319.

20   Ibid., p. 319.

21   Babbage, *On the Economy of Machinery and Manufactures*, p. iii.

22   Ibid., p. 175.

23   L.F. Menabrea, 'Sketch of the Analytical Engine Invented by Charles Babbage, with Notes upon the Memoir by the Translator, Ada Augusta, Countess of Lovelace', *Scientific Memoirs*, 1843, 3: 666–731, p. 706.

24   Charles Babbage, *Ninth Bridgewater Treatise: A Fragment* (London: John Murray, 1837), pp. 34–41.

25   Simon Schaffer, 'Babbage's Intelligence: Calculating Engines and the Factory System'.

26   Thomas Simmons Mackintosh, *The Electrical Theory of the Universe* (London: 1841).

27   Morus, *Shocking Bodies: Life, Death and Electricity in Victorian England* (Stroud: History Press, 2011).

28   'Horrible Phenomena – Galvanism!' *The Times*, 11 February 1819, p. 3.

29   Canto 1, Stanza 130, *Don Juan* (London: Penguin Books, 2004).

30   Quoted in Dorothy Stein, *Ada: A Life and a Legacy* (Cambridge MA: MIT Press, 1985), p. 22.

31   Quoted in Betty Alexandra Toole (ed.), *Ada, the Enchantress of Numbers: A Selection From the Letters of Lord Byron's Daughter and her Description of the First Computer* (Mill Valley CA: Strawberry Press, 1992), p. 53.

32   Quoted in Stein, *Ada: A Life and a Legacy*, p. 43.

33   Quoted in Toole, *Ada, the Enchantress of Numbers*, p. 51.

34   L.F. Menabrea, 'Sketch of the Analytical Engine Invented by Charles Babbage', p. 690.

35   'Report of the Committee Appointed to Consider the Advisability and to Estimate the Cost of Constructing Mr Babbage's Analytical Engine', *Annual Report of the British Association for the Advancement of Science*, 1879, 48: 92–102, p. 101.

36   H.W. Buxton, *Memoir of the Life and Labours of the Late Charles Babbage* (Cambridge MA: Charles Babbage Institute, 1988), p. 48.

37   Victor Horsley, 'Description of the Brain of Mr Charles Babbage, FRS', *Philosophical Transactions of the Royal Society, Series B*, 1909, 200: 117–31.

38   The story is reprinted in Brian Stableford (ed.), *Scientific Romance: An International Anthology of Pioneering Science Fiction* (New York: Dover Publications, 2017), pp. 69–86.

39   William Cooke Taylor, 'Objects and Advantages of Statistical Science', *Foreign Quarterly Review*, 1835, 16: 205–29, p. 207.

40   Quoted in Jack Morrell and Arnold Thackray, *Gentlemen of Science: Early Years of the British Association for the Advancement of Science* (Oxford: Oxford University Press, 1981), p. 292.

41   'Introduction', *Journal of the Statistical Society of London*, 1839, 1: 1–5, p. 2.

42   'Report of the Committee Appointed to Consider the Advisability and to Estimate the Cost of Constructing Mr Babbage's Analytical Engine', p. 100.

43   Charles Vernon Boys, 'The Comptometer', *Nature*, 1901, 64: 265–8, p. 266; p. 268.

44   United States Patent US395782A, 'Art of Compiling Statistics', p. 3.

45   Herman Hollerith, 'The Electrical Tabulating Machine', *Journal of the Royal Statistical Society*, 1894, 57: 678–89, p. 681.

46 Quoted in James Cortada, *Before the Computer: IBM, NCR, Burroughs and Remington Rand and the Industry they Created, 1865–1956* (Princeton: Princeton University Press, 1993), p. 51.

47 'Smee and Wiglesworth on Physical Biology', *British and Foreign Medico-Chirurgical Review*, 1849, 4: 371–82, p. 376.

48 'Instinct and Reason', *British and Foreign Medico-Chirurgical Review*, 1850, 6: 522–4, p. 524.

49 Jerome K. Jerome, 'Novel Notes', *The Idler*, 174–92, p. 188.

50 Alice Fuller, 'A Wife Manufactured to Order', *The Arena*, 13 July 1895, pp. 305–12.

51 United States Patent No. 200, 521, 'Improvement in Phonograph or Speaking Machines'.

52 Tesla, *My Inventions and Other Writings* (Harmondsworth: Penguin, 2011), p. 120.

53 Ibid., p. 121.

54 Ibid., p. 123.

Chapter 8: Flying High

1 Quoted in J.L. Pritchard, *Sir George Cayley: The Inventor of the Aeroplane* (London: Max Parrish, 1961), p. 206.

2 George Cayley, 'Sir George Cayley's Governable Parachutes', *Mechanics' Magazine*, 1852, 57: 241–4, p. 243.

3 Quoted in Pritchard, *Sir George Cayley*, p. 206.

4 'Henson's Letters Patent', *London Gazette*, 19 November 1842, p. 3295.

5 'Henson's Aerial Carriage', *Mechanics' Magazine*, 1843, 38: 258–63, p. 258.

6 'On the Aerial Steam Carriage', *Illustrated London News*, 1 April 1843, 233–4, p. 233.

7 Pritchard, *Sir George Cayley*, p. 153.

8 Letter from Cayley to Henson, 14 October 1846, quoted in Pritchard, op. cit., note 1, p. 190.

9 Pritchard, *Sir George Cayley*, p. 129.

10 'Astounding News', *New York Sun*, 13 April 1844.

11 William Bland, *The Atmotic Ship* (Sydney: David Mason, 1866).

12 Mark Twain, 'How is your Avitor?', *Alta California*, 1 August 1869.

13 'The Problems of Flight', *Bradford Observer*, 15 October 1868, p. 6.

14 'Crystal Palace', *Standard*, 6 July 1868, p. 3.

15 George Cayley, 'On Aerial Navigation', *Nicholson's Journal of Natural Philosophy*, 1809.

16 John William Strutt, 'On the Mechanical Principles of Flight', *Memoirs of the Manchester Literary & Philosophical Society*, 1900, 44: 1–26.

17 Tony Royle, *The Flying Mathematicians of World War I* (Montreal: McGill University Press, 2020).

18 Octave Chanute, *Progress in Flying Machines* (New York: American Engineer and Railroad Journal, 1894), Preface.

19 'Americans seem to have Solved Problem of Aerial Flight', *Newark Advocate*, 28 December 1903.

20 Quoted in Alfred Gollin, *No Longer an Island: Britain and the Wright Brothers, 1902–1909* (London: Heinemann, 1984), p. 193.

21 David Mindell, *War, Technology, and Experience Aboard the USS* Monitor (Baltimore: Johns Hopkins University Press, 2000).

22 Edward Reed, *Our Iron-Clad Ships: Their Qualities, Performances, and Cost* (London: John Murray, 1869), p. vi, p. x, p. xii, p. xvii.

23 H.G. Wells, 'The Land Ironclads', *The Strand Magazine*, 1903, 26: 751–69.

24 Frederick Jane, *All the World's Fighting Ships* (London: Little, Brown & Co., 1898).

25 George Griffith, *The Angel of the Revolution* (London: Tower Publishing Company, 1894), p. 72.

26 Ibid., p. 73.

27 Ibid., p. 82.

28 'Address Delivered at the Presentation of the Gold Medal of the Society to Mr Warren De la Rue', *Monthly Notices of the Royal Astronomical Society*, 1862, 22: 131–9, p. 133.

29 Ibid., p. 135.

30 Ibid., p. 136.

31 Warren De la Rue, 'Report on the Present State of Celestial Photography in England', *Report of the British Association for the Advancement of Science*, 1859, 29: 130–53, p. 143.

32 William Herschel, 'Astronomical Observations Relating to the Mountains of the Moon', *Philosophical Transactions*, 1780, 70: 507–21, p. 508.

33 Thomas Dick, *The Philosophy of a Future State* (New York: Carvill, 1829), p. 101.

34 Dick, *The Sidereal Heavens and Other Subjects Connected with Astronomy* (New York: Harper & Brothers, 1844), p. 208.

35 Robert Chambers, *Vestiges of the Natural History of Creation* (London: John Murray, 1844), p. 161.

36 Quoted in Michael Crowe, *The Extraterrestrial Life Debate, 1750–1900* (Cambridge: Cambridge University Press, 1986), p. 275.

37 Ibid., p. 282.

38 Ibid., p. 311.

39 William Rutter Dawes, 'On the Planet Mars', *Monthly Notices of the Royal Astronomical Society*, 1865, 25: 225–8, p. 226.

40 Ibid., p. 228.

41 Percival Lowell, *Mars* (Boston: Houghton, Mifflin & Co., 1896), p. 131.

42 Ibid., p. 135.

43 Agnes Clerke, *A Popular History of Astronomy during the Nineteenth Century* (London: Adam & Charles Black, 1893), p. 314.

44 Thomas Webb, *Celestial Objects for Common Telescopes* (London: Longmans, Green & Co., 1893), p. 164.

45 Ibid., p. 166.

46 Ibid., p. 188.

47 Maxwell, *Scientific Papers of James Clerk Maxwell*, p. 322.

48 H.G. Wells, *The First Men in the Moon* (London: George Newnes, 1901), p. 42.

49 Edwin Pallander, *Across the Zodiac* (London: Digby, Long & Co., 1896), p. 66.

50 John Jacob Astor, *A Journey in Other Worlds* (New York: D. Appleton, 1894), p. 6.

51 Ibid., pp. 89–90.

52 Griffith, *Stories of Other Worlds and Honeymoon in Space*, originally published in 1901 (Somerville MA: Heliograph Inc, 2000), p. 24.

53 Ibid., p. 72.

54 George Cayley, 'On the Principles of Aerial Navigation', *Mechanics' Magazine*, 1843, 38: 274–8, p. 278.

Index

Page references in *italics* indicate images.